Canadian Conservation Hall of Fame Inductee

THE SOIL FIXERS

Land stewards committed to the cause

My journey on the road to sustainable agriculture
examining the critical role of the earth's fragile soil

Harold B. Rudy
Ontario Soil
and Crop
Improvement
Association

Proceeds from this book belong to the non-profit Ontario Soil and Crop Improvement Association. Earnings will support student scholarships and on-farm soil and crop research.

 FriesenPress

Suite 300 - 990 Fort St
Victoria, BC, V8V 3K2
Canada

www.friesenpress.com

Copyright © 2018 by Harold B. Rudy
First Edition — 2018

Editor: Barry Gunn, Guelph, Ontario, Canada
Foreword: Dan Needles, Author and Playright

Materials contained in this publication are the property of Ontario Soil and Crop Improvement Association. Content may be used for private study and research. Published works must clearly acknowledge this publication. All rights reserved.

All rights reserved.

No part of this publication may be reproduced in any form, or by any means, electronic or mechanical, including photocopying, recording, or any information browsing, storage, or retrieval system, without permission in writing from FriesenPress.

ISBN
978-1-5255-2990-0 (Hardcover)
978-1-5255-2991-7 (Paperback)
978-1-5255-2992-4 (eBook)

1. TECHNOLOGY & ENGINEERING, AGRICULTURE, AGRONOMY, SOIL SCIENCE

Distributed to the trade by The Ingram Book Company

About the Author

Having grown up on a Mennonite family farm in Waterloo County, Ontario, Canada, as one of ten siblings, Harold Rudy is totally engrossed in all things agriculture. With a B. Sc. (Agriculture) and a M. Sc. (Rural Planning and Development), both from the University of Guelph, he has been privileged to dedicate his thirty-year career toward advancing the environmental stewardship programs of the Ontario Soil and Crop Improvement Association. As a 2017 inductee into the Canadian Conservation Hall of Fame and recipient of the Canadian Seed Growers' Association Life-time Achievement Award in 2016, Harold is solidly grounded in soil and crop matters and recognized as an authority on the many challenges confronting food production, environmental stewardship, and the ever-increasing pressures of our farm communities. He lives with his partner Sandra on the family homestead established by his parents in 1929. Nearly ninety years later, it continues to specialize in horticultural production.

The author's first accidental selfie, taken during a crop tour on Wolfe Island, Ontario, in 2011, was intended to photograph the farm's energy-producing windmills, which are only vaguely visible in the background.

Foreword

A HANDFUL OF DIRT

I learned in public school that many of the world's ancient civilizations perished because they lost their ability to feed themselves. The great empires of the Egyptians, Persians, Greeks, and Romans all declined and disappeared. Their monuments were covered by blow-sand because soil health remained a mystery to them.

If I had looked out the window of my little schoolhouse in 1958, I might have seen some evidence that we came perilously close to the same fate right there in Dufferin County. Between first settlement and 1900, the pioneers removed just about every tree from those sand hills and watched the topsoil wash away in torrents down the Nottawasaga River to Georgian Bay. During the Depression, many farmers walked away from their farms and the land reverted to the County for unpaid taxes.

My first job was tree-planting on those blow-sand farms. As a seven-year-old, I trotted behind two old men who had created Mono Township's first reforestation committee in 1925. Between the two of them, they had already planted a quarter of a million trees and they spoke with religious fervour about soil heath. They pointed to the great gullies in the hillsides and told me that I was walking chest deep in vanished topsoil. They insisted that it took poor land to make a good farmer.

This book captures over thirty years from the history of an organization that has been working for seventy-five years to arrest the decline of soil health in this province, a battle that is still far from being won. Harold Rudy guides us through the transition from local grassroots networks to an internationally recognized organization for soil advocacy. We meet a veritable who's who of Ontario agriculture as they struggle with competing agendas for best practices: conventional, no-till, organic, GMO, and anti-GMO. Along the way, they stumble on one good idea after another and draw on the talents of farmers, academics, civil servants, and public interest groups from every walk of life in the province.

The Soil Fixers reminds us that whatever our convictions might be about the best way forward for food production, there is no better way to bridge our differences than to get down to soil level and examine a handful of dirt together. Nothing less than our future depends on it.

Dan Needles is a writer who created the Wingfield Farm series of stage plays. He lives on a farm in Collingwood, Ontario.

Why I Became an Author

This book documents some of the highlights of my thirty-year journey to discover the true definition of sustainable agriculture and the critical role of soil. Chronicles about the Ontario Soil and Crop Improvement Association (OSCIA), my employer, have been written before. The first, written by former OSCIA secretary manager Arthur H. Martin, covered the period from its formation in 1939 until 1972. The second, Barbara Dyszuk's *Two Blades of Grass Where There Was One Before*, covered the years from 1972 to 1989.

My career with OSCIA has spanned the period from 1987 to 2018. Much of my tenure has focused on programs to facilitate testing of innovative equipment, trying new methods, and testing ideas or theories to enable those closest to the land — our farmers — to find their own path to sustainability. Thirty years has resulted in hundreds of boxed files, thousands of photos, scores of projects, hundreds of million dollars expended, and a multitude of tales waiting to be told.

I never had ambitions to be an author. Upon reflecting on a satisfying, and by most accounts, successful, career in this third era of OSCIA's history, it seemed to be a natural fit for me to tackle. I have been, after all, fully engaged in the happenings during this most recent time frame.

But there are additional compelling reasons for my story beyond capturing an organization's history. Soil and agriculture are inextricably linked to the carbon cycle and climate change. Farms can contribute to the climate crisis through poor soil management that releases carbon dioxide into the atmosphere. On the other hand, farmers can help fight climate change with soil-friendly practices that include no-till planting, planting cover crops, reducing soil movement and adding organic matter rich in carbon, all of which take in carbon dioxide and fix carbon in the soil. Growing crops successfully depends on good soil stewardship. Clean water depends on it. It is an important story to be told.

Assembling the story of *The Soil Fixers* was challenging. My stories and reflections are not just for those in farming. Those of my generation within the agriculture and food industry will be somewhat familiar with the chapter titles and contents. But if you are a few generations removed from the land yet interested in reading about food security, efforts to restore the land, fix our soil, tackle climate change, and protect our natural resources while we strive to pass on our farms to the next generations, *The Soil Fixers* is for you too.

My voyage provides glimpses of the soil managers and their best practices, all important ingredients for sustainable food production. Yet the word "sustainable" itself is seldom used throughout the text until the final chapters where I integrate the many ingredients of the environmental, economic, and social pillars of the sustainability framework established in the 1987 United Nations report, *Our Common Future*. The task of soil fixing is not yet complete. It is a continuous improvement process, as research provides new insights and as our land caretakers apply suitable methods to each of their unique farms. OSCIA's life story is not complete. There will no doubt be a fourth edition covering our next thirty years. I can't wait to read it!

Table of Contents

About the Author v

Foreword vii
    A handful of dirt

Why I Became an Author ix

Glossary of Acronyms xvii

Introduction xxi
    Partnerships, Programs, and
    Moving Forward

Chapter 1 1
    Soil: The Great Connector

Chapter 2 13
    Launch of Program Delivery

Chapter 3 31
    The Retirement Years: National
    Soil Conservation and Permanent
    Cover Programs

Chapter 4 43
    Who Got High on the High Crop
    Residue Program?

Chapter 5 61
    Who Sets the Agenda for Agriculture?

Chapter 6 67
    A World-Class Environmental
    Farm Plan

Chapter 7 101
       Look Out Below Your Feet

Chapter 8 107
       The Case of the Missing Barn Owl

Chapter 9 119
       What Do Crops Eat for Breakfast?

Chapter 10 133
       OSCIA's Evolving Role under Federal/
       Provincial/Territorial Frameworks

Chapter 11 145
       Modus Operandi

Chapter 12 177
       On-Farm Research and Collaboration

Chapter 13 233
       The Real Meaning of Sustainable Farms
       and Food

Conclusions 251

Acknowledgements 259

Appendix 1 263
       OSCIA Projects and Programs Listing

Appendix 2 270
       OSCIA Past-Presidents with Year Served
       and Home County/District

Appendix 3 274
       Confidentiality Agreement for
       Environmental Farm Plans

Appendix 4.............................276
   Report to the Ontario Federation
   of Agriculture
   Executive and
   Environmental Committees

Appendix 5.............................279
   Australian EMS Study Tour

Appendix 6.............................282
   Ontario Farm Groundwater Quality
   Survey 1991-1992

Appendix 7.............................283
   Past Ontario Forage Masters

Appendix 8.............................284
   OSCIA Projects Funded by Ontario
   Agricultural Adaptation Council

Appendix 9.............................286

Appendix 10............................291
   Summary of OSCIA Research
   Projects, 2006

Appendix 11............................293
   EFP First Edition Workbook –
   January 1994

Glossary of Acronyms

AAFC – Agriculture and Agri-Food Canada
AAC – Agricultural Adaptation Council
ADF – Acid Detergent Fibre
AGCARE – Agricultural Groups Concerned About Resources and the Environment
AGWEC – Agricultural Weather Centre
AFGC – American Forage and Grasslands Conference
APF – Agriculture Policy Framework
BMP – Best Management Practice
CAP – Canadian Agricultural Partnership
CSPS – Cellulosic Sugar Producers Co-operative
CFFO – Christian Farmers Federation of Ontario
CFIA – Canadian Food Inspection Agency
COFSP – Canada-Ontario Farm Stewardship Program
COFS – Canada's Outdoor Farm Show
CSA – Community Supported (or Shared) Agriculture
CSGA – Canadian Seed Growers' Association
EBI – Environmental Benefit Index
ENBP – Environmental Best Management Practices
EFP – Environmental Farm Plan
EMS – Environmental Management System
FAO – Food and Agriculture Organization
GE – Genetically Engineered

GHGMP –	Greenhouse Gas Mitigation Program for Canadian Agriculture
GMO –	Genetically Modified Organism
IFAO –	Innovative Farmers Association of Ontario
IMAC –	Interim Maximum Acceptable Concentration
IPM –	Integrated Pest Management, or International Plowing Match
LMAP –	Land Management Assistance Program
LSP –	Land Stewardship Program
GPS –	Global Positioning System
GF –	Growing Forward
GRCA –	Grand River Conservation Authority
GYFP –	Growing Your Farm Profits
MAC –	Maximum Acceptable Concentration
MLW –	Multi-level well
MNR –	Ministry of Natural Resources
MOEE –	Ministry of Environment and Energy
NCC –	Nature Conservancy Canada
NDF –	Neutral Detergent Fibre
NMS & NMP –	Nutrient Management Strategy and Nutrient Management Plan
NSCP –	National Soil Conservation Program
OAC –	Ontario Agricultural College
OFA –	Ontario Federation of Agriculture
OFEC –	Ontario Farm Environmental Coalition
OMAFRA (OMAF) –	Ontario Ministry of Agriculture, Food and Rural Affairs
OSCEPAP –	Ontario Soil Conservation and Environmental Protection Assistance Program
OSCIA –	Ontario Soil and Crop Improvement Association

OSGA –	Ontario Seed Growers' Association
PGRC –	Plant Gene Resource Canada
PMRA –	Pest Management Regulatory Agency
PNT –	Plants with Novel Traits
PLUARG –	Pollution from Land Use Activities Reference Group
RUSLEFAC –	Revised Universal Soil Loss Equation for Application in Canada
SAR –	Species at Risk
SARPAL –	Species at Risk Partnership on Agricultural Land
SCCC –	Soil Conservation Council of Canada
SFFI –	Sustainable Farm and Food Initiative
UN –	United Nations

Introduction

PARTNERSHIPS, PROGRAMS, AND MOVING FORWARD

I have been fortunate to have worked with exceptional colleagues, partners, and farmers during my career at the Ontario Soil and Crop Improvement Association (OSCIA). I can list at least two-dozen hallmark programs created and delivered during my tenure that have had a significant influence on creating awareness, motivating change, encouraging adoption of improved land stewardship practices, and providing financial incentives to help the farm community over the financial hurdles of investing in change.

In Ontario alone, over $200 million has been invested in the farm community through program contracts over the past thirty years. This investment, mostly from the provincial and federal governments through the Ontario Ministry of Agriculture, Food and Rural Affairs (OMAFRA) and Agriculture and Agri-Food Canada (AAFC), has leveraged another $400 million invested by the farm community in land stewardship activities.

Behind these many contracts are stories that need to be told, and I've had a great vantage point for watching them unfold. If not in the middle of the fray, I've at least been on the sidelines as an observer. There are at least a half-dozen significant milestones — real game changers — that have kept OSCIA relevant, responsive, and visible in the quest for better land stewardship.

OSCIA has played a key part in numerous initiatives that have had a major impact over the past several decades, including the:

- Land Stewardship Program (LSP)
- National Soil Conservation Program (NSCP)
- High Crop Residue program
- Environmental Farm Plan (EFP) program
- Greenhouse Gas Mitigation Program for Canadian Agriculture (GHGMP)
- Nutrient Management Financial Assistance Program
- Accommodating Species at Risk
- Growing Forward
- Growing Your Farm Profits (GYFP) program
- Sustainable Farm and Food Initiative (SFFI)

These are just a few of the programs that I will describe in the following pages. OSCIA and the agricultural community benefited from many others as well. A full listing of over one hundred programs and projects since the 1980s can be found in Appendix 1.

OSCIA also focuses on the improvement of crops themselves. It researches better methods for farmers to deal with the risks of weather, insects, disease, weeds, and seed variety selection (resulting in improved quality and yield) so they can eke out a slightly better return on investment. Farmers are always striving for improvement—always fixing things!

While program funding for addressing environmental concerns has come easily to OSCIA, it has been more challenging to acquire funding for field activities related to increasing crop yields. At the local level, cultivating strong working relationships with the extension staff in OMAFRA's Field Crops Unit has been an integral part of addressing the crop production needs of our local members and growers in the broader farm community. A key feature of OSCIA's work that sets it apart from commercial research is the impartial on-farm field trials carried out under the watchful eye of the government extension staff. As a senior manager, one of my greatest challenges has been acquiring and maintaining resources to support research for crop production, even though these efforts have been recognized by our members as a high priority. This challenge will no doubt continue for managers in the years to come.

Co-operation and Communication

> "The challenge is not to do more but to communicate what we are already doing. The OSCIA is involved in many different field plots, activities and committees where the major challenge is to give members and directors access to this information and to keep them feeling involved in the process. Regional newsletters, blogs, websites and teleconferences were developed to fill this need. In my time as director, the provincial board went from meeting twice a year to 10-12 teleconferences per year, now to potentially daily updates with social media. Communication continues to be most important for our organization."
>
> - Joan McKinlay, 2012 OSCIA President (Grey County)

OSCIA's communication network has also matured over the past thirty years. Eleven regional associations have been formed to encompass the fifty-three county/district associations[1] that are affiliated under the constitution of the OSCIA. On-farm research results are published through *Crop Advances*, an annual summary of OSCIA's field trials. Local workshops, annual meetings, and field days provide opportunities to profile major accomplishments in soil and crop research, meeting the needs of members and the surrounding farm community. The county/district associations are masters at organizing these local events, and social media is expanding opportunities to connect with members and the broader farm community. Over thirty years, I have worked for over thirty OSCIA presidents. How many people do you know who have worked for thirty bosses? The names of these dedicated leaders, as well as honorary presidents, are listed in Appendix 2.

---

1 Most sub-provincial jurisdictions in Ontario are counties or regional municipalities, but in Northern Ontario, they are called districts.

Farming Systems

Farming methods affect soil health, of course, and there are controversies surrounding farming methods. Not the least of these have been triggered by the very scientific advancements that the farm community relies upon for continual improvement and economies of scale. Genetic engineering, for instance, has been the result of scientific research, though it is the subject of controversy as well, and OSCIA has not been immune to debate involving scientific advancements. What are the concerns of our members? How have genetically engineered crops allowed farmers to reduce or eliminate soil erosion with reduced tillage? Soil tillage is one of the most destructive practices affecting soil health and, in this book, I'll tell you why.

Figure 1.01 – Poor resilience to soil erosion in the spring of 2018 – far too common. What Best Management Practices (BMPs) should be adopted here? (Photo source: OSCIA)

Partnerships, Programs, and Moving Forward

Figure 1.02 – High resilience to soil erosion in the spring of 2018. Soil protected with lots of crop leaves and stems with a history of conservation tillage, alfalfa, wheat, and cover crop rotation with livestock manure application. This farming system will survive for centuries. (Photo source: OSCIA)

Where does organic agriculture fit in? Are organic farming systems more sustainable than non-organic? I've included a section entitled "The Organic Paradox" that shows how demand for organic food exceeds supply. Why aren't more farmers adopting organic methods? Where are the blurry lines between non-organic and organic? I have visited numerous organic farms in my travels throughout the Great Lakes Basin. My observations may surprise you.

The terms "corporate farming" and "family farm" also evoke strong feelings. Are family farms more responsible soil managers? I have been the beneficiary of heart-warming exposure to cream-of-the-crop farm families. Over 97 per cent of Canadian farms are family owned.[2]

Many of our family farms are indeed incorporated for a bunch of reasons, but we don't use the term "corporate farms" in this book. Nor do we use the term "industrial farms." When does a farm become industrial? What is a non-industrial farm? This is partly because when the media use the term "industrial," they often portray a company with absentee investors. But when it comes to farms in Ontario, that image does not match the reality. I have observed a multitude of

---

2   Statistics Canada. 2016 Census of Agriculture. www.statcan.gc.ca/daily-quotidien/170510/dq170510a-eng.htm (accessed August 8, 2017)

farming philosophies touted by a wide range of farming practitioners dedicated to their families, their farms, their surrounding communities, their soil management, and the environment. We need them all, but assuredly, these farming systems will continue to strive for continuous improvement.

This book is not just a history of the dozens of program contracts that OSCIA has administered. It is also a reflection of challenges faced by food producers as they strive to satisfy shifting consumer tastes and growing demands from broader society wanting to know more about how we farm. Yes, there are many opinions about how we should farm. With the world's population expected to approach ten billion within the next generation, increasing demand for food will continue to challenge farmers and test the resilience of our soil, the thin layer of Earth's crust that is essential for the survival of our civilization.

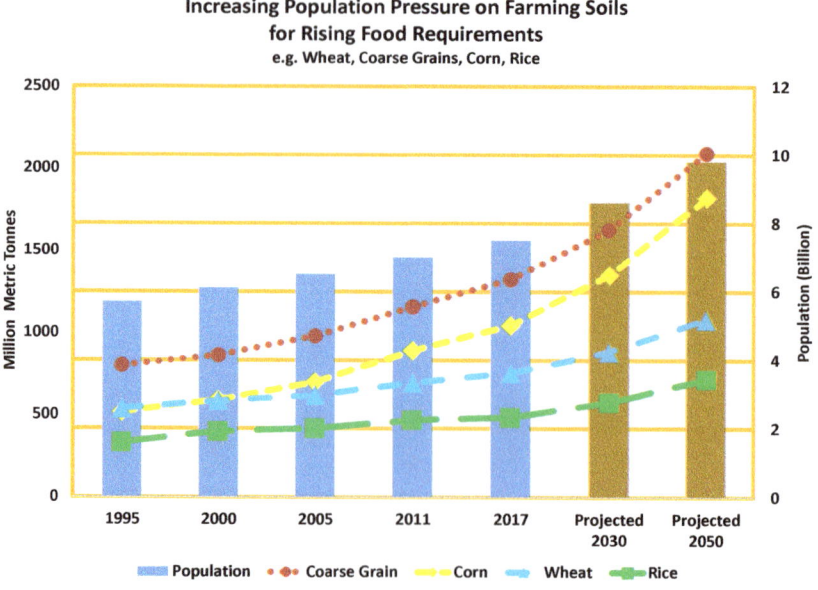

Figure 1.03 - The rising demand and supply for food increases strain on our fragile soils.[3]

---

3   Chart compiled by OSCIA with data from Foreign Agricultural Service, USDA Office of Global Analysis, https://apps.fas.usda.gov/psdonline/circulars/grain.pdf and World Meters, http://www.worldometers.info/world-population/#table-historical accessed January 30, 2018)

Please join with me in a reflection on OSCIA's past thirty years. I worked under thirty-two OSCIA presidents, all farmers, and all totally committed to the betterment of OSCIA and Ontario agriculture. This book is a tribute to them, our OSCIA members, and to our colleagues and partners. If you enjoy food, I hope you'll be inspired by their efforts and dedication to feeding the world while saving the environment. Protecting our agricultural soils and the surrounding landscape is a major priority.

Along the way, you'll learn about the Soil Champion Award. I'll explain who got high on the High Crop Residue program, and how Ontario finally received its provincial soil designation (called the Guelph Soil Series) during the United Nations International Year of Soils in 2015. Although the Year of Soils has come and gone, our soils live on, providing cleaner air, food, and filtered water when they are managed with care, innovation, and thought for the well-being of future generations.

Chapter 1

SOIL: THE GREAT CONNECTOR

"We're responsible for feeding those people," I said to the woman in the seat beside me as we looked out the window from high above Atlanta down to the thousands of lights below. That city had over five million people to feed, and it was only one of huge numbers of large, growing cities requiring food around the world. I'm not in the habit of blurting out off-the-wall comments to strangers about feeding the masses, but my fellow traveller had asked me where I was coming from, where I worked, and why I was heading to Toronto. Before I had my seatbelt fastened, I had given her enough detail about my career in agriculture, food, and environmental matters that she knew I had a keen interest in food production—feeding people. She also understood that "we" meant the agriculture and food industry, and we were responsible for feeding not just the five million or so in Atlanta, but people everywhere.

I was returning from a soil conservation conference in Omaha, Neb., and realized more than ever that feeding a growing population places increasing pressure on the existing land base—the soil that farmers steward. Soil—that layer of the earth's crust affectionately known by most as the stuff we want in our gardens and flower beds. Farmers want it too! Rich, fertile, and water-holding, a teaspoon of the richest soils have more living microorganisms than there are people on Earth. These living factories surround plant roots to provide moisture and nutrition to the plants.[4]

---

4   United States Department of Agriculture [USDA], *Soil Health Nuggets: There Are Some Amazing Things Going on Underground*, retrieved from U.S. Department of Agriculture website www.nrcs.usda.gov/Internet/FSE_DOCUMENTS/stelprdb1101660.pdf (accessed January 10, 2017).

What Is Soil?

When I googled the question, "what is soil?" 11.2 million results came up. Not surprising, as soil is the great connector. Except for seafood, greenhouse vegetables grown in artificial media, and mushrooms grown on oak logs, soil is responsible for producing the food on our tables, feed for livestock and poultry, and a host of other vegetation. In suitable environments and with the right ingredients, it grows trees and plants that provide countless benefits, including habitat and food for wildlife. Plants absorb sunlight, and through the "mini-factories" responsible for photosynthesis, the vegetation sucks up carbon dioxide to produce the oxygen that we need to survive.

Soil is also a sponge that absorbs surplus rainfall. When working correctly, it slowly releases that water to growing plants or percolates into streams, rivers, and underground aquifers. It acts as a filter that prevents contaminants and nasty germs from getting into our water supply—the aquifers deep underground and the rivers, lakes, and streams on the earth's surface. I attend conferences where chemists and chemical engineers tell farmers that in the future, targeted crop species and biomass crops will replace fossil fuels as a source of energy; hence the growing importance of keeping our soils productive. Soil produces vegetation used to manufacture everything from car parts to the clothes on our backs.

Here's the real shocker about the importance of good soil management: at least fourteen per cent of global greenhouse gas (GHG) emissions come from agriculture, according to the Organisation for Economic Co-operation and Development (OECD).[5] However, as one OECD report indicates, agriculture can play a major role in reducing GHG emissions through improved nutrient and soil management, work that the OSCIA and its partners have been doing for the past few decades.

In addition to all the good work that OSCIA has done over the years, we are continually striving to add to our knowledge base, and that is why I went

---

5   Based on OECD (2010), Climate change and agriculture, OECD Observer n°278, March 2010
    http://oecdobserver.org/news/archivestory.php/aid/3213/Climate_change_and_agriculture.html. (March 2010)

to Omaha to attend the annual Mid-West Cover Crops Council Conference. While I was there, I added to the knowledge and experience I'd gained during my thirty-year career of delivering programs to conserve, protect, and improve soil. This included learning about how to use cover crops to protect soil during the winter and spring months when it's most vulnerable to damage. As a senior manager for OSCIA, I have encountered continual challenges as my colleagues and I have designed programs to motivate and assist Ontario landowners to be the best stewards possible. These are urgent challenges, not just for Ontario farmers and their counterparts in the United States, but also for all land managers around the world. Much has been accomplished, but there is much yet to do.

> Civilizations have been destroyed because they neglected their soil

We're the soil fixers—always tinkering to make soil better. Like an automobile, soil requires maintenance and sometimes repairs. As a living organism, it requires fuel to keep functioning. It also requires remanufacturing, enhancements, investment, and conservation, not just to serve immediate needs, but also to preserve this valuable resource for future generations.

When I meet people not involved in farming and tell them that I'm employed by the Ontario Soil and Crop Improvement Association, a puzzled look often crosses their faces, as they don't completely understand what the "improvement" part entails. Does it involve pesticides, organic production, or genetically modified organisms (several of the most commonly debated agricultural topics in the social media)? Rather than engaging in debate, I often divert the discussion, and that's what I did with my seatmate on the plane. I just said that I'd attended the conference in Omaha to learn more about soil conservation and that soil improvement is a complicated task that is sometimes ignored.

Civilizations have been destroyed because they neglected their soil. In his book titled *Dirt: The Erosion of Civilizations*, Dr. David R. Montgomery describes how the fall of the Mesopotamian, Greek, Roman, and many other ancient civilizations was caused, in large part, by mismanagement of the land. Hillsides eroded into irrigation canals, rendering the up-slopes unproductive for food and choking the lowland canals with upland silt. This, in turn, reduced or eliminated the water supply required for irrigation. These patterns were evident across the ancient world, including North Africa, the breadbasket of the Roman Empire.[6] Dr. Montgomery presented these same ideas to delegates at our annual meeting in London, Ont., in 2015. Most likely, the survival of my children, my grandchildren, and their children will depend on critical decisions relating to the soil conservation practices of our generation.

When I started my career, the terms "sustainability" and "sustainable farming" were seldom used. It is a much different story today as the OSCIA and its partners work to address sustainability issues through working groups such as the Sustainable Farm and Food Initiative (SFFI).[7] OSCIA is also a member of the Canadian Roundtable for Sustainable Crops, which brings together agriculture and food organizations several times per year from across Canada.[8]

Looking back at the programs and projects delivered by OSCIA, all were important precursors to the way that today's agriculture and food industry responds to growing demands for sustainable sourcing. OSCIA, along with other industry partners, is continuing to build on the foundation of over seventy-five years of stewardship practices.

Soil is central to OSCIA's history, but it is also part of the wider landscape. Typically, that landscape is what members of the non-farm community see in their travels through rural communities. Yes, they notice crops growing in fields, but they also observe woodlots, wetlands, wildlife, a mix of fencerows with shrubs and bushes, and of course, farm buildings, homesteads, silos, and grain bins. OSCIA is involved with these too. In fact, promoting and enhancing biodiversity in the landscape as a whole has been an important niche for OSCIA.

---

6 Montgomery, D.R. *Dirt: The Erosion of Civilizations* (Berkeley and Los Angeles: University of California Press, 2007).

7 Sustainable Farm and Food Initiative [SFFI], website www.sustainablefarms.ca.

8 Canadian Roundtable for Sustainable Crops, website www.sustainablecrops.ca.

This book is not a technical guide to soil management. Many other publications deal with chemical, physical, and biological recipes for good soil stewardship. Yet the descriptions of programs in the ensuing chapters will profile how farmers seek to adopt best soil management practices.

What's the Big Deal about Saving Soil?

> *"Every organization, including OMAFRA and AAFC, has identified soil health as a major concern. OSCIA is recognized as being at the forefront of many soil health initiatives."*
>
> - Gord Green, 2016 OSCIA President (Oxford County)

When I started my career, from time to time one would see a car or truck sporting a bumper sticker that said "Save Our Soil." But concerns about preserving this valuable resource have been around for longer than that. The United States formed the Natural Resources Conservation Service (NRCS) in 1935 to address the devastating soil erosion of the Dirty Thirties,[9] and the Government of Canada established a federal department called the Prairie Farm Rehabilitation Administration (PFRA) to address western Canadian wind erosion and drought concerns.[10]

Then in 1939, a group of dedicated Ontario farmers set out on a mission to improve their cropping practices and improve soil management. They'd just been through the Dirty Thirties and vividly remembered the Dust Bowl years on the Canadian Prairies and the Great Plains of the United States. They formed the Ontario Crop Improvement Association as a network of farmers to assist scientists at the Ontario Agricultural College (OAC) and government extension staff (experts and scientists assigned to outreach activities) from what was then known as the Ontario Department of Agriculture. In 1952, the word "soil" was added to

---

9   USDA. www.nrcs.usda.gov/wps/portal/nrcs/detail/national/about/history/?cid=nrcs143_021392)

10 Gilson, J.C. (December 16, 2013). "Prairie Farm Rehabilitation Administration," The Canadian Encyclopedia www.thecanadianencyclopedia.ca/en/article/prairie-farm-rehabilitation-administration/.

complete the name as we know it today: the Ontario Soil and Crop Improvement Association.[11]

OSCIA's mission is "To facilitate responsible economic management of soil, water, air, and crops through development and communication of innovative farming practices."[12] With 4,000 members, OSCIA is a significant presence in all the major agricultural areas in Ontario. Members are actively seeking, testing, and adopting optimal farm production and stewardship practices. Our longstanding development and delivery of stewardship programs, supported by the provincial and federal governments, provide outstanding educational and cost-share funding opportunities for all producers.[13] We've been at it since 1939, and I've been in the middle of it for the past thirty years.

> Dr. Lobb estimates that soil erosion costs Canadian agriculture about $3 billion in lost profits each year

In the mid-1980s, prior to my employment at OSCIA, I worked as a soil conservation adviser for what is now the Ontario Ministry of Agriculture, Food, and Rural Affairs (OMAFRA). I was part of the joint conservation program with the Grand River Conservation Authority (GRCA) and my office was located at the GRCA's headquarters in Cambridge, Ont. My employment partner was a bright and budding summer student out of University of Toronto, but more importantly, an experienced farm lad from Huron County. We trucked and towed conservation equipment up and down the Grand River watershed to make it available to farmers on a trial basis, so they could experiment with new conservation measures.

That student was David Lobb, now a professor of landscape ecology and senior research chair in watershed systems research at the University of Manitoba in Winnipeg. He is one of Canada's preeminent soil scientists.

---

11   OSCIA (Spring 2016) "Handbook for Local and Regional Associations," retrieved from OSCIA website www.ontariosoilcrop.org/wp-content/uploads/2015/10/1.SecretaryHandbookSpring2016.compressed.pdf, accessed April 5, 2017.

12   Ibid.

13   ibid.

Dr. Lobb estimates that soil erosion costs Canadian agriculture about $3 billion in lost profits each year. In a paper titled "Soil Degradation: The Cost to Agriculture and the Economy," he presented his recent findings to the Summit on Canadian Soil Health, held in Guelph, Ont., in August 2017 and sponsored by the Soil Conservation Council of Canada.

"This loss is attributed to about 10 per cent of the total thirty-nine million hectares of cropland in Canada which has remained moderately to severely eroded over the past thirty years," he says. "Based on the 2011 Census of Agriculture, it works out to about $80 per hectare of cropped land, on average. Obviously, the loss is much greater per hectare on moderately or severely eroded cropland. And, this works out to about $30,000 per year (on the moderately to severe eroded cropland) per farm, on average[14] ... From the most recent national assessments, it is not clear that soil conservation measures have stabilized these losses; in some regions, these losses continue to increase in severity and extent."[15]

One source estimates the global cost of land degradation at up to US$10 trillion per year.[16]

With that in mind, it was a fitting move when the United Nations declared 2015 as the International Year of Soils. According to the UN's Food and Agriculture Organization (FAO), "Healthy soils are the basis for healthy food production; soils are the foundation for vegetation which is cultivated or managed for feed, fibre, fuel and medicinal products; soils support our planet's biodiversity and they host a quarter of the total (biodiversity); soils help to combat and adapt

---

14   Lobb, D. A., "Soil Degradation: The cost to agriculture and the economy." https://www.researchgate.net/publication/319701531, accessed December 19, 2017)

15   Lobb, David A. (Nov. 19, 2016) "The Cost of Soil Erosion and Sedimentation to Canadians and the Impact on Water Bodies Across Canada: An Overview Prepared for The Soil Conservation Council of Canada." https://www.researchgate.net/publication/311054526_The_Cost_of_Soil_Erosion_and_Sedimentation_to_Canadians.

16   One World News. "Land Degradation Cost Put at Up to $10 trillion A Year." oneworld.org., The Economics of Land Degradation, www.oneworld.org/2015/09/10/land-degradation-cost-put-at-up-to-10-trillion-a-year (Accessed April 5, 2017).

to climate change by playing a key role in the carbon cycle; soils store and filter water, improving our resilience to floods and droughts; soil is a non-renewable resource, its preservation is essential for food security and our sustainable future."[17]

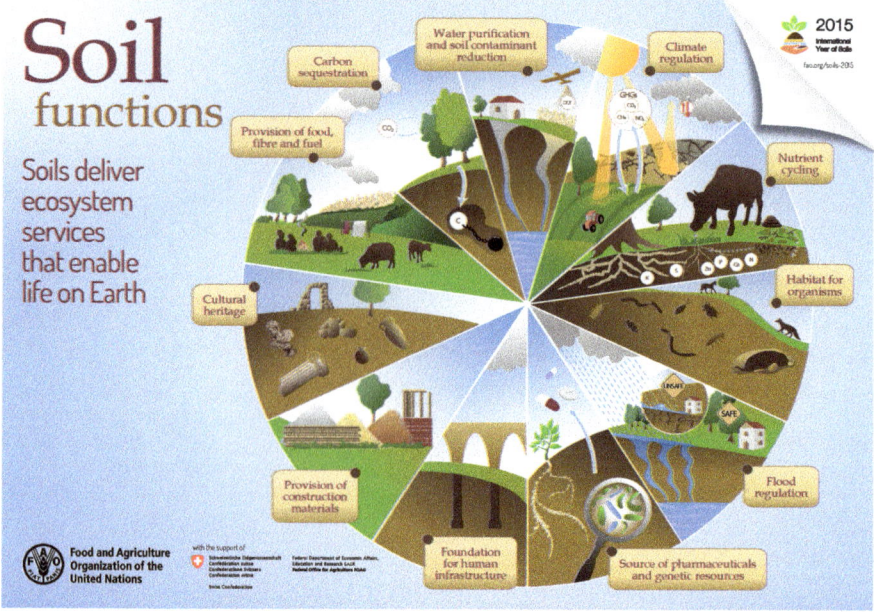

Figure 1.1 The diagram provided by the FAO illustrates the many ways that healthy soil contributes to life on Earth.[18]

The UN declaration was followed by the Vienna Soil Declaration by the International Union of Soil Sciences (IUSS), which declared 2015-2024 as the International Decade of Soils.[19]

---

17   Food and Agriculture Organization of the United Nations, "Healthy Soil for Healthy Life," www.fao.org/soils-2015/about/key-messages/en/ (Accessed March 23, 2017)

18   Food and Agriculture Organization. Soil Functions, www.fao.org/resources/infographics/infographics-details/en/c/284478/ (Accessed February 20, 2018). Reproduced with permission.

19   Vienna Soil Declaration, Soil matters for humans and ecosystems, http://iuss.boku.ac.at/files/vienna_soil_declaration_december_7_.pdf (Accessed January 25, 2018)

Soil is essential to our survival, and that, folks, is why we need soil fixers! They make soils healthy and keep them that way. They restore degraded fields that have been damaged by unusual storms, improper management, or lack of knowledge.

Taking lessons from previous lost civilizations, we know that productive, healthy, resilient soil is a priority not just for our children and grandchildren, but for generations to come.

Much has been accomplished in the past thirty years and OSCIA continues with this work. However, problems still haunt us, the soil managers, not just in Ontario but around the world. I continue to work with researchers, experts from government and agribusiness, and farmers, to fix some of the most important challenges facing our planet.

There are many dedicated to being soil fixers, and that was the purpose of the Omaha conference, where like-minded professionals assembled to learn and share the latest on soil management. Why, I wondered on the flight home, were there still devastating problems due to soil mismanagement, soil erosion, and the associated water pollution not only in Ontario, but elsewhere as well? The agenda for the conference in Omaha included numerous soil fixers profiling their successes. Why weren't there more?

What Is Healthy Soil?

There is plenty of healthy soil in Ontario, yet OSCIA members and many other interested stakeholders are constantly looking for improvement. We are continually identifying areas where soil needs some fixing. It may be a portion of a field — a knoll, a steep side slope, a whole field, or even a whole farm. Perhaps it is just an unproductive corner near a wetland that should be shifted into a productive wetland for ducks or aquatic invertebrates.

Ideally, we want the activity of earthworms, microbes, and other soil-dwelling creatures that we can't normally see with the naked eye to improve the soil for us, and we are learning more and more each year about the value of those small critters. If I were to emphasize one takeaway lesson about soils, I would want everyone to understand the importance of these living organisms and how they produce healthy plants above the soil.

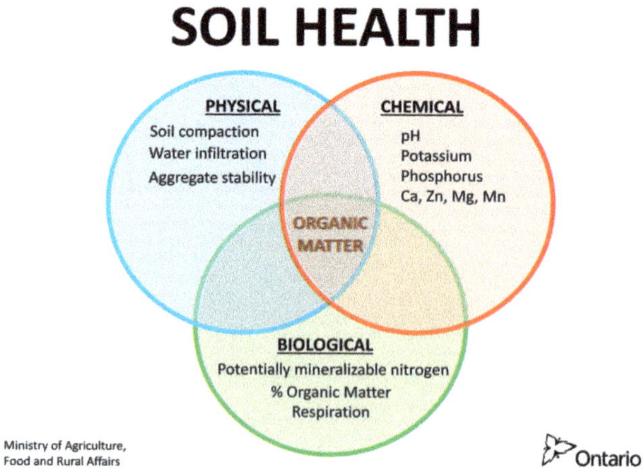

SOIL HEALTH

PHYSICAL
Soil compaction
Water infiltration
Aggregate stability

CHEMICAL
pH
Potassium
Phosphorus
Ca, Zn, Mg, Mn

ORGANIC MATTER

BIOLOGICAL
Potentially mineralizable nitrogen
% Organic Matter
Respiration

Ministry of Agriculture, Food and Rural Affairs

Ontario

Figure 1.2 Components of soil health.[20]

Here's what the soil experts from OMAFRA have to say on the subject. Healthy soils: Provide minimum resistance to root growth, improved crop development, and ultimately high yields and product quality;

- Provide better returns on crop inputs such as applied nutrients and pesticides;

- Allow for better infiltration, more water storage, and less runoff;

- Are more resilient during low water conditions because their structure and organic matter content help retain plant-available moisture;

- Resist degradation, such as compaction, crusting, water and wind erosion, and ponding;

- Are better equipped to remove pollutants and protect groundwater quality; and

---

20   "Sustaining Ontario's Agricultural Soils: Towards a Shared Vision." http://www.omafra.gov.on.ca/english/landuse/soil-paper.pdf, 8 (Accessed March 23, 2017)

- Reduce emissions of greenhouse gases including carbon dioxide, methane, and nitrous oxide. Implementing Best Management Practices (BMPs) for soil health—especially those that add organic matter—will improve the soil's ability to serve as carbon and nitrogen sinks.[21]

Why Does Healthy Water Depend on Healthy Soil?

In soil science terms, healthy soils require good "aggregates." In other words, we want the soil particles to hang together (resist falling apart), but leave just the right amount of space between them to allow air movement, water absorption, space for roots to move freely, and biological organisms to flourish. Good aggregate stability contains just the right ratio of minerals, organic matter, pore space, microflora and fauna, readily available nutrients, and moisture. Where soil aggregates are poor, water does not absorb well and becomes contaminated by silt when soil particles are unstable and fall apart. See Figure 1.3 below as an illustration of the Slake Test to determine the soil health, or compare levels of aggregate stability.

OMAFRA recommends that we can improve a soil's aggregate stability by:

- Avoiding tillage practices and any activities that disturb a soil to break up existing aggregates, breakdown organic matter, and prevent accumulation of organic matter;
- Keeping the soil covered. Uncovered and unprotected soil is exposed to the physical impacts of raindrops and wind;
- Adding organic amendments regularly; and
- Avoiding pesticides harmful to beneficial soil microorganisms.[22]

---

21   OMAFRA. "Best Management Practices: Soil Health in Ontario." (brochure) www.omafra.gov.on.ca/english/environment/bmp/series.htm#1

22   OMAFRA. "Aggregate Stability." http://www.omafra.gov.on.ca/english/crops/hort/news/hortmatt/2015/07hrt15a2.htm (Accessed January 11, 2018).

Figure 1.3 The Slake Test illustrates the health of soils.
(Photo source: OMAFRA)

The Slake Test is a great way to check the condition of soil. OMAFRA soil management specialist Adam Hayes (Figure 1.3) demonstrates how chunks of soil from different management systems react when they are dried and placed in water. When the soil is placed on the wire baskets, the poorer managed soils are easily broken apart, clouding the water. The cloudy water represents poor aggregate stability, while the clear water represents good aggregate stability — a condition for which to strive. The sample with little soil left in the basket comes from continuous soybeans in conventional tillage. The others represent improving systems for crop rotation; the clearest sample on the left is from a three-year crop rotation with no-till management. In 2014, Hayes became the first recipient of OSCIA's Soil Champion award in recognition of his twenty-five years of service to agriculture.

We have briefly reflected on how soil is the great connector and basis for our survival. Future chapters will provide a glimpse of many determined soil fixers dedicated to the managing their farms for generations to come.

Chapter 2

LAUNCH OF PROGRAM DELIVERY

"Do you have any questions for us?" asked Richard Sovereign, then-president of OSCIA, in October 1987. There is nothing unusual about that question in a job interview, but I was dumbfounded, as it was the first question he asked after I entered the room and slid into the waiting chair. I had applied for the position of program manager to head up a new initiative called the Land Stewardship Program (LSP). That question usually comes at the end of an interview, so I was caught off guard and initially speechless.

A dozen potential responses flashed through my mind. Certainly, questions such as "How much ya paying?" or "Will my office have a window?" were not appropriate, nor were they asked. Somehow, I mumbled a query about how the working relationship would be structured between OSCIA and the Ontario Ministry of Agriculture and Food (OMAF, as OMAFRA was known then). Little did I know at the time that this would be one of the most relevant questions I could ask. The partnership between OSCIA and the ministry was highly valued, and the historically strong relationship between the two organizations helped give credibility to the proposed three-year LSP partnership.

In any case, I was able to satisfy Richard and the other three vice-presidents about my qualifications, and I was offered the job. Years later, I learned that I was not everyone's first choice. I was a rookie, with no experience in industry leadership like other candidates. Apparently, it was a tie and one of the vice-presidents had to break it. It was simple, he told me. Experienced candidates wanted more compensation than I did, so adhering to the frugal nature of the farm community, the tie-breaker voted for me because I'd requested a lower salary.

Program Delivery: A New Direction for OSCIA

> *"While on the Executive in the late 1980s, it seemed like we were pioneers, changing the future of OSCIA with the formation of the first Land Stewardship Program."*
>
> *-Jim Yungblut, 1990 OSCIA President (Niagara North Region)*

I was launched into my new career, and uncharted territory, with a bang! Jack Riddell, then Ontario's agriculture minister, announced the $40-million LSP at the International Plowing Match in September 1987 at the John Lowe family farm near Meaford in Grey County. By October, Ontario farmers were champing at the bit, like a horse heading for home. "Where can I apply for my land stewardship grant?" callers would ask as they flooded the phone lines to government offices across Ontario.

It was a struggle for me to accept my new job offer. I was becoming established as one of OMAF's soil conservation advisers, a position which appeared to be evolving into long-term job security. Why would I leave a secure government position for a three-year contract filled with uncertainty? I discussed this dilemma with Galen Driver, senior manager of the soil and water management branch, who suggested that if I did accept this three-year management job, I would become more employable. Galen was right! Of course, my employability had as much to do with new and growing opportunities opening up over the next thirty years at OSCIA. I sent off a letter of resignation by snail mail to my boss, Jim Arnold, who was on a special assignment in Sri Lanka at the time. He acknowledged my choice (again by snail mail) and my new career began.

When I started with OSCIA as a program manager on Nov. 1, 1987, the LSP was the first of its kind in Ontario, with a multimillion-dollar contribution provided by OMAF. OSCIA signed a contract to administer $31.3 million of that $40-million provincial commitment. We were in charge of motivating landowners to adopt conservation tillage, supported by cost-share cheques for eligible projects that were dispersed among thousands of farming landowners to make improvements to soil and water protection.

Launch of Program Delivery

Before 1987, OSCIA had been a low-key farm organization with barely a few thousand dollars in the bank, but it did have fifty-three local associations across Ontario. As a progressive network of farmers concerned about everything to do with soil and crops, OSCIA was well-positioned for grassroots program delivery. For nearly fifty years, OSCIA had been focusing on field days, crop tours, local workshops and on-farm demonstrations of new tillage, plant variety comparisons, and cropping methods, so it seemed totally logical to engage the association to improve land stewardship efforts. It was determined that stakeholders should be engaged in decision-making and administration—a new model for program delivery. Committees would be set up in each county/district across Ontario to administer a portion of the funds locally, and in the process, they would coax neighbouring farmers to adopt improved land stewardship practices. Funds were allocated proportionately based on the number of row-crop acres in each area. Row-crop acres include corn, soybeans, coloured beans, cereals, and vegetables but do not include hay or pasture. The acreage data was obtained from Statistics Canada.

> As a progressive network of farmers concerned about everything to do with soil and crops, OSCIA was well-positioned for grassroots program delivery

Scientific studies carried out during the 1970s had determined that the Great Lakes were at risk because of soil erosion and degenerating water quality. For where soil moves into water courses, so too does phosphorus, a potent nutrient that stimulates plant growth. Phosphorus is good when it's in the field next to the roots of a valuable plant, but bad when it stimulates the growth of algae and water-based plants in nearby water bodies. The more that water-based algae thrives, the more oxygen they extract from the water, leading to poor water quality. The Great Lakes, Chesapeake Bay, the Mississippi River watershed, and now Lake Winnipeg are high-profile water bodies threatened by eutrophication, a process that disrupts the ecological balance when prolific algae die off and decay, using up the dissolved oxygen in the water.

The nutrients that algae love come partly from agriculture (including livestock manure and commercial fertilizer), but water bodies also receive nutrients from urbanization, sewage treatment plants, recreation areas, homeowner lawns, and inadequate septic systems on cottage and rural properties. My job was to focus on farm sources of these nutrients. Others would have to deal with the non-farming contributors.

There was full justification for serious action in the years leading up to the announcement of the LSP in 1987. Numerous briefings had been submitted to government bodies about the devastating erosion of Ontario's crop land. For years, scientists had been studying the effects of soil erosion on water quality in the Great Lakes. In 1978, a team of scientists who were part of the Pollution from Land Use Activities Reference Group (PLUARG) released a report that outlined the devastating effects of silt and phosphorus, which were stimulating prolific growth in aquatic plants.[23] By the time the 1980s arrived, no one could remember what PLUARG stood for, but we did know that there were serious water-quality problems in the Great Lakes. People on both sides of the Great Lakes were demanding action through agreements signed between the Canadian and U.S. governments.

No-Till Farming

"OSCIA has always worked with other groups with great success for soil and crop improvement."

-Elwin Vince, 1992 OSCIA President (Kent County)

Meanwhile, back at the farm, soil scientists and astute farmers were taking notice as they observed their topsoil washing downslope, leaving degraded rills and gullies in their fields. This, of course, reduced crop yields. Much of the erosion was caused by disturbing the soil by plowing in the fall, leaving the soil bare to the erosive forces of nature over the winter and following spring (before the newly planted crops could provide shelter with their canopy and stabilization from their roots). At the time, it was often noted

---

[23] International Reference Group on Great Lakes Pollution from Land Use Activities (PLUARG). "Environmental Strategy for the Great Lakes System: Final Report to the International Joint Commission." agrienvarchive.ca/pluarg/pluarg_final_report.html (Accessed March 23, 2017)

that farmers were doing excessive tillage in the spring to prepare a seedbed. The term "recreational tillage" was often bandied about.

In the early 1980s, a delegation from the Ontario government, including agricultural engineer Jim Arnold and field crop specialists John Schleihauf and Neil Moore, headed to Ohio to learn first-hand how American farmers were using no-till methods to prevent soil erosion and maintain crop yields. According to Mr. Arnold, there was great doubt about whether the no-till concept would work in Ontario, where temperatures were cooler and soils higher in moisture later into the spring. In fact, the debate was so intense among this threesome that on the return trip to Canada, the driver—Mr. Arnold—was so engaged in the discussion that he did not initially acknowledge the customs agent who patiently waited in his kiosk, leaning on his chin, until he finally asked what the argument was all about. Mr. Arnold simply reported that they were having a debate about soil conservation. The border agent looked at them indulgently, acknowledged the government logo on the side of their car, and said, "Okay, go ahead." I wonder how things would turn out today if one approached a border crossing and ignored the agent while getting in one last verbal jab at your colleagues as the border police look on.

Coincidentally, in 1980, an astute and innovative Huron County farmer by the name of Don Lobb (father of Dr. David Lobb) became one of the first farmers in Ontario to successfully adopt the aforementioned no-till concept of planting that challenged conventional thinking. Mr. Lobb was also an OSCIA member and a passionate soil conservationist.

No-till planting is a method of planting crops with minimal disturbance of the soil. No-till farmers do not plow at all, neither in the fall nor in the spring. Planters or seeders insert seeds into the soil in a slot made by a disc opener. Sometimes, if there is too much debris from the previous year, additional trash removers may be used to clear a six- to eight-inch pathway through the debris. The soil surface remains covered with straw, stalks, and plant debris on the surface from the previous crop. The surface cover and undisturbed roots protect the soil through the winter when the snow melt and devastating rains would otherwise cause bare soil to wash away. Plant residue left on the soil surface also acts as small dams or filters that hold water back, allowing the water to infiltrate the soil more slowly instead of rushing across the surface and becoming an erosive force. Also, when soil is not tilled, better soil aggregates can develop. This is essential to reduce water and wind erosion, for water infiltration, and for reduced compaction.

But farmers had to be convinced that leaving a messy-looking field covered in stalks and straw would not threaten their crops. Mr. Lobb participated in the 1983 Soil Today – Food Tomorrow conference organized by OSCIA, the University of Guelph, and OMAF, where he explained his experience to other farmers. He also organized the first Conservation Tillage Demonstration at the 1984 International Plowing Match, held at the J.D. Ross farm near Teviotdale in Wellington County.[24]

Mr. Lobb's farm also became a case study for a Senate of Canada report called *Soil at Risk: Canada's Eroding Future*. He also gave numerous lectures in Ontario and internationally describing how to successfully use conservation practices. This coincided with a growing movement of farmers who wanted to learn how to be better soil managers. Without question, his efforts helped initiate the Land Stewardship Program, (LSP), which raised awareness of the issues and provided cost-share funding for farmers to assist them in their transition to improving soil conservation.

OMAF's Joint Conservation Programs, Conservation Authority Assistance, and the Land Stewardship Program (LSP)

> *"There was concern when OMAF withdrew (provincial) secretarial support but the conversation soon turned to opportunity where the connection to OMAF was maintained by hiring Doug Wagner as secretary-manager to run the day-to-day operations with direction from the executive. I believe this was the start of the biggest opportunity the OSCIA has ever had, culminating with the first Land Stewardship Program."*
>
> -Maurice Martin, 1991 OSCIA President (Elgin County)

---

24   Home of the International Plowing Match and Rural Expo, Past Sites, https://www.plowingmatch.org/about-us/archives/past-match-sites (Accessed January 30, 2018).

Launch of Program Delivery

I began my work in soil and water conservation as a part-time employee of OMAF through the winter of 1984. My job was a clerical one, in which I screened cost-share applications for farmers. These applications were funded by the province under the Ontario Soil Conservation and Environmental Protection Assistance Program (OSCEPAP). OSCEPAP grants were designed to support enhancements to erosion control structures such as grass waterways, tile outlet protection, inlet filters, and other capital improvements. They did not include funding for conservation tillage or cropping methods. The next generation of the program, OSCEPAP II, provided funding for manure storages and improved management of manure. I was hired by Jim Arnold, assistant manager of the soil and water management branch.

Figure 2.1 OMAF's team of soil conservation advisers, circa 1988
Back row, L-R: Jack Kyle, Rob Templeman, Neil Moore, Chris Attema, Adam Hayes, Peter Roberts, Doug Aspinall, Ted Taylor, Andy Graham; Middle row, L-R: Peter van Adrichem, Mark Janiec, Brent Kennedy, Pierre-Yves Gasser, Dan Wright, Graham Gambles, Terry Davidson, Steve Clark, Wilf Shier, Don Dietrich; Front, L-R: Marianne vanden Heuvel, Anne Verhallen, Lisa Cruickshank, Howard Lang, Helen Lamers-Helps, Sheila Nolan, Christine Brown (Yours truly had moved on to a new employment opportunity with OSCIA)
(Photo source: Original Archived by Andrew Graham, OSCIA

The initial thrust for soil conservation occurred when Dennis Timbrell, Ontario's agriculture minister from 1982 to 1985, announced an increase in the OMAF budget to hire over twenty soil conservation advisers. This put the wheels in motion for Jim Arnold and Galen Driver to expand field staff across Ontario and work with regional conservation authorities for the common good of keeping soil on farms, and out of our rivers and lakes. The conservation authorities became valuable partners in the Joint Conservation Program because they had the capacity to store, transport and operate conservation farm equipment. OMAF purchased no-till drills and other conservation equipment that was loaned to farmers for experimentation in their fields comparing no-till management with traditional cropping methods. The conservation authorities had repair shops so it was logical to store and transport equipment from their yards. The conservation equipment demonstrated to farmers how they could better manage their soils, while maintaining crop yields. Until the introduction of the LSP, cost-share grants were not available for conservation tillage. LSP was the missing link that joined the expertise of OMAF soil conservation advisers and cost-share funding to help producers transition to improved soil management.

With its badly needed focus on soil and cropping management, the LSP broadened opportunities for farmers. The program was championed in cabinet by Jack Riddell, the agriculture and food minister, and there was reference to funding for a land stewardship program in the spring 1987 Speech from the Throne. OSCIA moved quickly by drafting and submitting a proposal to OMAF on June 30, 1987.

Doug Wagner, OSCIA's secretary-manager, led the charge, recruiting Janet Horner from OMAF to wordsmith the proposal. In fact, Doug, Janet, then-OSCIA president Richard Sovereign, and Ken Knox, a senior manager of OMAF, met at a hotel in Mississauga, Ont., that spring to brainstorm about this position paper and to determine how the proposal should be structured. It has been reported that neither the ministry nor OSCIA had funds to buy lunch!

The Christian Farmers Federation of Ontario (CFFO) also submitted a proposal to the ministry outlining the need and their support for improving land stewardship. Support from the CFFO complemented OSCIA's proposal.

However, there was one problem. Immediately after the LSP was announced to the farm community in September 1987, the phones started ringing off the hook. No detailed program guide had yet been developed. There were no application forms and no organized and trained staff available to assist with the paperwork. In my circles, it was well known that staff at OMAF were already fully engaged with other duties. New employees would have to be hired and trained to deliver the LSP. OSCIA and OMAF scrambled to piece together a delivery model featuring the following key elements:

- It would be modelled after a similar program operated successfully by the Tobacco Marketing Board, in which local producers play a key role in developing, implementing, and administering the program.

- Program guidelines and policies would be developed by a board consisting of representatives from OMAF and OSCIA, with input from other producer organizations.

Administration would be handled by a small paid staff employed by OSCIA.[25]

---

25   OSCIA. "Proposal for Land Stewardship Program" (June 30, 1987, p. 12). OSCIA Archives. The following text formed the basis for the delivery model:

- "The Ontario Soil and Crop Improvement Association ask (sic) that the Ministry consider operating the Land Stewardship Program in partnership with the OSCIA. The delivery could be modelled after the current program administered by the Tobacco Marketing Board. In this way, a producer organization would take an active role in the development, implementation and administration of the Land Stewardship Program.

- Program guidelines and policy would be developed by a policy board consisting of Branch Directors from the Ministry and appointed representatives of OSCIA. Input from other producer organizations and agencies would be sought to ensure that [the] program complemented existing federal and provincial programs and … [was] designed for maximum benefit. Directors from the Soil and Water Branch, Plant Industry and Agricultural Representatives are suggested as members of the policy board. Cheques will be issued by the policy group or designate.

- Administration would be handled by a small paid staff employed by OSCIA. An 800 number could direct inquiries to this group. Consultations, bidding, application processing, and program approvals would be handled by contract and casual professionals. In many cases, it will be preferable to employ as many as 250 farmers on a casual basis to assist in program delivery."

The "small paid staff" is where I came in as the program manager. Teri Bryce Cobean was hired as the accounting manager, and Janice Murray Kyle was brought on as administrative assistant. OMAF gave us free office space, which at that time, was surplus space rented at the Ignatius Jesuit Centre just north of Guelph on Highway 6. Perhaps this was divine intervention for OSCIA.

By coincidence or celestial foresight, a roundtable discussion was organized at OSCIA's annual meeting in February 1987 to hear grassroots suggestions about how OSCIA could become more self-sufficient, rather than continue to rely on annual government grants. This discussion led to a motion calling for the OSCIA executive to approach the agri-business community for financial support.[26] Allan Yungblut, a delegate from the Niagara North Association, brought the motion forward, and it was passed by the delegates. This set the wheels in motion for the OSCIA to pursue program delivery opportunities. The LSP was the first.

Components of the Land Stewardship Program

According to OMAF's 1987–88 program brochure, the LSP was a "three-year, $40-million program to provide financial incentives for first-time adoption of conservation farming practices on Ontario farmland." It was designed to "enhance and sustain agricultural production and improve soil resources and water management by: reducing soil erosion and soil compaction; restoring soil organic matter and structure; and minimizing potential for environmental contamination from agricultural practices."

The LSP consisted of four components:

1. Financial assistance grants for:

    a.) Soil building and maintenance

    b.) Erosion control structures

    c.) Conservation machinery and equipment

    d.) Technical training

2. Research related to stewardship practices

---

26 OSCIA. Minutes of the Annual General Meeting. Constellation Hotel, Toronto. (Feb. 3-4, 1987)

3. Education and extension
4. Program delivery and services[27]

Training the Local Land Stewardship Program Committees

A key feature of OSCIA's proposal was to train local committees of farmers in each county/district across Ontario to guide and direct program delivery. A local field person was hired for field inspection, administrative duties, and promotion of the LSP opportunities. These field inspectors were the contacts who would answer questions, administer the application forms, inspect completed projects, and forward claims to head office for payment.

By the time I came on board as manager of LSP for OSCIA, most of these committees had already been appointed. The plan was to have a few key senior ministry managers and several OSCIA executive members fan out across Ontario to "train" these committees and their field staff at regional training meetings. Different individuals were assigned to different regions of Ontario to spread out the workload.

A framework had been provided, but detailed program guidelines had not yet been fully developed. As a result, each region received slightly different instructions. There was a general philosophy that each local committee had significant latitude to administer the program in their jurisdiction based on the needs and conservation priorities of their area.

Have you ever given mixed messages to your kids? Engaging multiple trainers resulted in messages that were not entirely consistent, through no fault of the trainers. It was a big task to train 265 newbies in the fifty-three counties/districts of the province. A few committees felt they had total autonomy and authority to design the program without any consideration for provincial consistency. A few others felt that since livestock farmers were such good land stewards with their crop rotations, some of the funds should simply be given to them as a reward for past stewardship excellence on their hay and pasture lands (hay and pasture grasses are considered the best

---

27 OMAF. "Land Stewardship Program Description." (brochure) www.agrienvarchive.ca/lsp/lspbroch.html (Accessed on April 5, 2017).

vegetation for protecting and improving soil). "Why reward the sinners?" was a common sentiment expressed at meetings across the province with reference to crop farmers whose soil had eroded, which would make them eligible for LSP grants.

Not all ministry extension staff in the county/district offices were supportive of the revolutionary concept of putting farmers in charge of accepting or rejecting conservation proposals from their neighbours. We received the occasional finger-wagging comment, suggesting that this model would never work!

> Farmers, as grassroots decision makers, were empowered to institute meaningful change

Thankfully, open-minded committees of farmers saw things differently. Committee chairs like Eric Kaiser from Lennox-Addington County, for example, got up and defended this new concept. Delegating responsibility to farmers, he proclaimed confidently, should be given a chance. Eric was already a soil conservation advocate working with challenging Napanee clay soil near the Bay of Quinte. He was a former army captain, and when Eric spoke in his commanding bass voice, everyone listened. He was confident that many new conservation methods would be developed by innovative farmers.

Eric was right, and he made my job as a rookie manager so much easier. In fact, his support for the new program model was replicated across Ontario. Farmers, as grassroots decision makers, were empowered to institute meaningful change. In communities across Ontario, it was the talk of the town. Hockey games, coffee shops, equipment dealers, and even church meetings were often locales where conversation turned to LSP committee members and their neighbouring farmers. Would this or that project be eligible? What do you think about this concept? These interactions played out across Ontario so successfully that within the first year, all the available funds for grants were committed to 6,000 farmers. Other farmers were left on a waiting list.

The LSP Committees Face-to-Face

It was my job to promote the program across Ontario, clarify some of LSP's vagaries, and nudge a few dissenter committees toward a more consistent path. Over the first few winters, many speaking engagements involved lots of travel, early starts, and long days as I'd made a commitment to meet personally with all fifty-three county/district committees in their own communities. It took two years to achieve that goal. I typically arranged a breakfast or luncheon meeting with the full committee and the field inspector to discuss their challenges, hear of successes, and suggest how a stewardship program could be modified for the future.

Those first two years were among the most rewarding of my thirty-year career. Who were the people recruited for the LSP committees? I soon discovered that their backgrounds included much more than farming. Most had post-secondary education either at a college or a university. At least one, Bob Hart, had a master's degree—his being in soil science from the University of Guelph. In fact, Mr. Hart reminds me to this day that he delayed completion of his master's by a year because he was so busy as chair of the Oxford County LSP committee. Yikes! That must have cost him another year's tuition.

Many LSP committee members had experience in other vocations or careers before, or along with, farming. There were retired school teachers, retired bank managers, municipal councillors, mayors, retired police officers, firefighters, agribusiness salespeople, former government employees, and more. Some had international experience. By and large, they were active in their local communities as volunteers in coaching or firefighting or in other farm organizations or their church.

OSCIA was incredibly fortunate to have acquired these cream-of-the-crop, experienced, dedicated, and respected decision makers. We insisted that the committees consist of a cross-section of people from throughout each county/district and be representative their area's commodity mix (e.g., livestock, crops, horticulture). Many opened up their farm homes to host the four-person committees for meetings.

Richard Sovereign, the OSCIA president who hired me in 1987, advised me that one of my first tasks was to buy a portable cell phone. He had actually hired me over his cell while he was combining corn. Yes, he made an employment offer while talking to me over a $3,000 suitcase phone he carried with him in his combine cab. Cell phones were not common in 1987, but Richard was a progressive thinker and soon learned that a cell phone would save him an incredible amount of time when on the road, in the fields, or in the barn or workshop. I went out shopping as directed but cringed at the price tag. Eventually, I settled for a $2,000 model that was a fixed unit for my Mazda pickup truck. Richard and I laugh to this day about me being hired over a $3,000 cell phone.

Lennox-Addington County LSP chair Eric Kaiser's prediction about the wisdom of empowering a local committee to disperse public funds among their fellow farmers proved to be true. Not only did the program sell out quickly, but there were few complaints from producers. The success of a government program is measured in part by the number of letters of complaint received by a cabinet minister's office, and it had been common for farmers to complain about the rules of previous programs. If a producer's application for funding was rejected by a government official, the producer might call their Member of Provincial Parliament, who in turn would submit an inquiry to the agriculture minister's office to find out about the details. But there were few complaints about the LSP.

The LSP committees resolved complaints locally and were not afraid to say no or decline a questionable project. It was said that they were tougher in approving projects than government employees, but that they were honest, transparent, and fair in their deliberations. The cost of administration was also very reasonable, at less than 10 per cent of the overall $30-million contract for the duration of the three-year agreement.

The overwhelming success of LSP has resulted in dozens of OSCIA-led program contracts after 1987, and program delivery is now entrenched in OSCIA's business plan. Many of these will be outlined in the pages that follow.

LSP: The Next Generation

> *"I was President for OSCIA's fiftieth anniversary year and instrumental in replacing the antique logo representing a sheaf of grain, into our modernized present logo."*
>
> -Bill Zandbergen, 1989 OSCIA President (Dundas County)

On the heels of the original program's success came a second generation of LSP: Land Stewardship II (LSII). LSII represented a continuing commitment to conservation systems and environmental protection measures that had been part of the original LSP. It was a $38-million program, effective from Sept. 1, 1990, to March 31, 1994, and it had four major aspects:

1. A new concept, identified as Conservation Farm Planning, required producers to complete an analysis of their farming practices.
2. Extension, education, and technology transfer field staff were made available to producers.
3. Grants were provided
    a.) for farmers who adopted practices or built structures as part of their Conservation Farm Plan;
    b.) for setting up organizations, which would perform on-farm demonstrations and evaluation; and
    c.) for organizations that were established for conservation promotion and education.
4. Administration was led by OSCIA to (again) establish local land stewardship committees to facilitate this administration. These committees reviewed conservation farm plans and projects for funding, and they hired part-time staff to assist with program administration.

Notably, LSII was part of the Canada-Ontario Accord for Soil and Water Conservation and Development. With this accord, Canada and Ontario agreed to co-ordinate soil and water conservation programs.[28]

LSII provided a comprehensive list of options allowing Ontario producers to request financial assistance to improve soil management and protect waterways, including:

- residue management to protect the soil surface from erosion;
- cover crops for additional soil improvement;

---

28 "Canada-Ontario Accord for Soil and Water Conservation and Development." (Oct. 12, 1989) www.agrienvarchive.ca/nscp/download/nscp_agree.pdf (Accessed on March 23, 2017).

- strip cropping, where a strip of hay crop was alternated with corn or soybeans to slow water movement on steep slopes;
- conservation equipment;
- erosion control structures to channel water in a protective manner;
- manure storage and handling systems, including wash water systems from dairies; and
- pesticide storage and handling facilities for environmental protection.[29]

It should be noted that the emphasis on Conservation Farm Planning in LSII led to that program becoming the forerunner of the successful Environmental Farm Plan (EFP), which is still available today.

How Successful Were the Land Stewardship Soil Fixers?

If there were tonnes of soil eroding downslope in farmers' fields, waterways filling up with sediment, and Great Lakes water quality being impacted by sediment and phosphorus, how much did LSP and LSII reduce these impacts? Our management committees did not discuss how we could measure this impact. Nor was any effort expended as part of these early programs to build in credible metrics.

Nonetheless, $2.25 million was available for research under LSP. This resulted in research that focused on very specific topics and geographical locations. However, none of this research provided big-picture insights into what was happening within the Great Lakes region as a whole.[30]

---

29   OMAF. "Land Stewardship II: 1990 - 1994" (brochure) http://agrienvarchive.ca/lsp/lsp2.html (Accessed on March 23, 2017).

30   National Soil Conservation Program in Ontario Final Report. (Ottawa: Agriculture and Agri-Food Canada). Appendix G, pp. 160-201. (Accessed on March 30, 2017).

To a certain extent, that work was underway elsewhere. The Great Lakes Water Quality Program[31] established that surface runoff was the primary pathway for silt and nutrients to travel into the Great Lakes. It also established methods for measuring nonpoint source pollutants, and evaluated remedial measures to reduce soil erosion and the impacts of runoff.

> "Tillage destroys and/or depletes the soil's aggregate stability, structure, pore space, water holding capacity, infiltration, permeability, gaseous exchange and nutrient storage ability."
> (John Graham, USDA)

Another major research project, the Soil and Water Environmental Enhancement Program (SWEEP), provided a hundred final reports, many of which are available online.[32]

LSP and LSII provided over $50 million to the farm community in the form of cost-share grants for on-farm improvements. Little thought was given in that era to include performance measures that could track improvements to the environment. However, scientists were starting to ask tough questions about our significant investment and what effect it was really having on soil and water quality. This was the beginning of performance anxiety, which is discussed more fully in a later chapter.

Today, many farmers and soil conservationists understand that too much tillage destroys soil health over time. John Graham, a soil health specialist with the U.S. Department of Agriculture's Natural Resources Conservation Service (NRCS) in Kentucky, put it this way:

---

31   Wall, G.J., D.R. Coote, C. DeKimpe, A.S. Hamill, F. Marks. "Great Lakes Water Quality Program: Overview of Program." Great Lakes Advisory Committee, Agriculture and Agri-Food Canada. (March 1994) http://agrienvarchive.ca/glwq/glwq1.html#SUMMARY%20OF%20ACHIEVEMENTS
(Accessed on March 17, 2017).

32   Bowman, Bruce. T. (archivist) Soil and Water Environmental Enhancement Program. http://agrienvarchive.ca/sweep/sweephom.html
(Accessed March 30, 2017)

"Tillage destroys and/or depletes the soil's aggregate stability, structure, pore space, water holding capacity, infiltration, permeability, gaseous exchange, and nutrient storage ability." [33]

LSP was a great beginning to empower many farmers with knowledge, financial incentives, and tools to transition their soil management practices away from the degrading historical traditions of plowing and cultivation used for centuries. Today, there are thousands of exceptional soil managers in Ontario, expanding their soil biodiversity. This improved soil health supports greater crop productivity while reducing any offsite environmental impact. But what about farm soils that were so badly degraded or at elevated risk that even conservation tillage wasn't a suitable cure? Perhaps there were fields or a portion of a field that should never have been cleared of trees for agriculture. The next chapter covers a new era of program delivery that broadened the options for astute soil fixers.

---

33   Graham, J. "Tillage Destroys Soil Physical Properties." USDA Natural Resources Conservation Service. https://www.nrcs.usda.gov/wps/portal/nrcs/detail/ky/soils/?cid=stelprdb1096792 (Accessed July 24, 2017)

Chapter 3

THE RETIREMENT YEARS: NATIONAL SOIL CONSERVATION AND PERMANENT COVER PROGRAMS

No, we did not go into retirement after the Land Stewardship Program (LSP). Even before the first generation of LSP expired, and based on the overwhelming success of local program delivery by OSCIA's counties/districts, the federal government was taking notice. Federal administrators were looking for answers after a less than satisfactory delivery of a drought assistance program that provided payouts based on rainfall calculations recorded at select metering locations. The recorded rainfall (or lack thereof) was extrapolated across a broader landscape. In theory, that should have worked fine and perhaps did in western Canada. But in Ontario, rainfall can be sporadic, hitting one farm heavily while a neighbour on the next concession receives nothing. Agriculture and Agri-Food Canada (AAFC), too, was looking for another model to deliver programs in Ontario.

In 1989, the National Soil Conservation Program (NSCP) was announced to address concerns of soil degradation across Canada. There was strong consensus by soil experts that some agricultural land should be retired from agricultural production. Fields that were too rolling, or those last few acres right up against a stream, were considered too fragile for row crops (grains, oilseed, or vegetables). It was anticipated that every farm could find an acre or two that was high risk to environmental degradation or not profitable because of degradation. The previous 1983-1990 Ontario Soil Conservation and Environmental Protection Assistance Program (OSCEPAP) provided

$50 per acre (as a one-time payment) for agricultural land retirement but there was little uptake. In western Canada, land retirement for similar reasons was underway and $50 per acre went much further across the Prairies than in Ontario. In the United States, the Conservation Reserve Program (CRP) under the 1985 Farm Bill paid farmers significant subsidies for ten- to fifteen-year land retirement commitments for marginal land. They received what was equivalent to rental payments of roughly $50 per acre per year. U.S. farmers applied for funding assistance under a bid system. At its peak, there were 36 million acres enrolled in this program.[34]

> Senator Herb Sparrow chaired the Senate Committee on Agriculture, Fisheries, and Forestry as it held hearings across Canada to investigate soil degradation

Keep in mind that in the 1980s, commodity prices were low, interest rates were high, and farmers were stinging from poor returns. With the mounting evidence of the impacts of soil erosion, there was pressure to take highly erodible land out of annual crops to plant permanent grass or trees. Over the years, the CRP evolved to support conservation practices including:

- Drinking water protection
- Reducing soil erosion
- Wildlife habitat preservation
- Preservation and restoration of forests and wetlands
- Aiding farmers whose farms are damaged by natural disasters[35]

---

34   USDA. "Conservation Spending Seeks to Improve Environmental Performance in Agriculture." www.ers.usda.gov/topics/natural-resources-environment/conservation-programs/background (Updated Oct. 17, 2016, accessed April 21, 2017)

35   USDA Farm Service Agency. "Conservation Programs." www.fsa.usda.gov/programs-and-services/conservation-programs/index (Accessed Feb. 27, 2017)

Soil Conservation Council of Canada

Canadian farmers looked across the U.S. border and wondered, "Why isn't a similar program available here?" The charge was taken up by Senator Herb Sparrow, a North Battleford, Sask., businessman and rancher who chaired the Senate Committee on Agriculture, Fisheries, and Forestry as it held hearings across Canada to investigate soil degradation. The committee consulted farmers, researchers, agencies, and organizations and produced a report entitled *Soil at Risk: Canada's Eroding Future*.

"Canada is facing the most serious agricultural crisis in its history and unless action is taken quickly, this country will lose a major portion of its agricultural capability. The Standing Senate Committee on Agriculture, Fisheries and Forestry has travelled extensively in Canada examining the issue of soil degradation, a problem which is already costing Canadian farmers more than $1 billion per year in farm income. It has determined that we are clearly in danger of squandering the very soil resource on which our agricultural industry depends."[36]

On behalf of the OSCIA, then-president Laurence Taylor submitted a brief to the committee on May 1, 1984. Here's another quote from the report: "Responses to date by government and the agricultural community have been out of scale with the magnitude and severity of the problem, which only threatens to worsen before it gets better. The need is urgent for a major, well-organized, and adequately funded response to soil erosion and soil degradation."[37]

Huron County farmer and OSCIA's Don Lobb presented his farm as a case study to the Senate committee. In Guelph, Ont., the heavy artillery came out as nearly two dozen agronomists, farmers, scientists, and conservationists voiced their observations and views in two days of testimony.[38] The group included:

---

36   Senate of Canada. "Soil At Risk: Canada's Eroding Future." http://www.albertasenator.ca/flashblocks/data/Soil%20at%20Risk/Soil%20at%20Risk.pdf. p. 1 (Accessed April 21, 2017)
37   Ibid. p. 8. (Accessed April 21, 2017)
38   Ibid., p. 89. (Accessed April 21, 2017)

The Ontario Soil and Crop Improvement Association: Mr. Laurence Taylor, President.

The Ontario Institute of Pedology: Mr. Galen Driver, Program Manager, Soil and Energy Management, Plant Industry Branch, Guelph Agriculture Centre.

The Ministry of Agriculture and Food of the Province of Ontario: Dr. Robert McLaughlin, Director, Plant Industry Branch, Guelph Agriculture Centre; Dr. Vernon Spencer, Director, Capital Improvements Branch.

The Rondeau Bay Watershed Agricultural Steering Committee: Mr. Jack A. Rigby, Chairman.

Ecologistics Limited: Mr. Dave Cressman, President.

Mr. Jim McGuigan, M.P.P. (Kent-Elgin).

University of Guelph: Mr. Willen van Vuuren, Professor, Department of Agricultural Economics and Extension Education.

The Ontario Hay Association: Mr. Fritz Trauttmansdorff, Vice-President.

The Ontario Institute of Agrologists: Mr. Paul Fish, Chairman, Soil Conservation Committee; Mr. Don McArthur, Executive Director.

The County of Oxford: Mr. Charles Tatham, Warden of the County of Oxford, Woodstock, Ontario.

The Association of Conservation Authorities of Ontario Subcommittee on Soil and Water Conservation: Mr. Dennis O'Grady, Agricultural Technician.

Mr. Charles Shelton, Ingersoll, Ontario.

The Thames River Implementation Committee: Mr. Art W. Bos, Agricultural Diffuse Source Control Program Co-ordinator.

The Soil Conservation Society of America, Ontario Chapter: Mr. Bryan D. Boyce, President; Dr. Charles S. Baldwin, Soil Erosion and Sedimentation Committee.

The Ontario Farm Drainage Association: Mr. Kenneth R. McCutcheon, President.

Mr. Heinz Kumpat, Kitchener, Ontario.

The evidence presented was clear and new co-ordination was required. This prompted the formation of the Soil Conservation Council of Canada (SCCC) with Senator Sparrow as its first chair. From today's SCCC website, a tribute is provided: "The Soil Conservation Council of Canada (SCCC) has a rich history as the face and voice of soil conservation in Canada. The SCCC was founded in 1987, by a group of individuals under the leadership of Senator Herb Sparrow, to advocate for the importance of soil conservation on a national scale."[39]

> "The Soil Conservation Council of Canada (SCCC) has a rich history as the face and voice of soil conservation in Canada…" (SCCC website)

Hitting the Ground Running in Ontario

This Senate of Canada committee's work was before my time at OSCIA but it set the stage for action across Canada to flag unsustainable soil management practices. The committee's report provided strong justification to launch a set-aside program in Canada, similar to what had been established in the U.S. With the establishment of the National Soil Conservation Program (NSCP), over $8 million was available in Ontario for on-farm land retirement projects. The offering had to be more lucrative than the previous offering of $50 per acre. Tree planting was seen as a viable alternative but subsidizing the cost of buying the trees alone would exceed $100 per acre.

A further complication challenged us at the launch of the NSCP. There were not sufficient tree stocks available to accommodate large-scale tree planting throughout the proposed three-year life of the NSCP. Tree nurseries had to seed and plant their stocks three to four years in advance before transplanting them on farms. Furthermore, all the NSCP funds had to be spent by the time the program ended, so the program design could not be set up like in the U.S. where producers continued to receive annual payments for ten to fifteen years as long they continued with their land retirement commitments. This was a major dilemma for OSCIA and its administrators.

---

39   Soil Conservation Council of Canada. "About Us." www.soilcc.ca/about-us.htm (Accessed April 21, 2017)

How could we design our program to wisely utilize the federal NSCP funds within a three-year window?

After considering the options—and consulting with partners from OMAF, the Ministry of Natural Resources (MNR), Ontario conservation authorities, and Agriculture and Agri-Food Canada (AAFC)—we elected to design the program with similar characteristics to the LSP. Each county/district in Ontario would be allocated a portion of the funds, and each of the local OSCIA committees was charged with distribution of their allocated amount to projects that met the goals of permanent cover. AAFC signed a contribution agreement with OSCIA in August 1990 to deliver the program.

To solve the dilemma of short-term funding for long-term commitment, OSCIA introduced a unique new concept. A bid system would allow producers to declare what they required as compensation for a land-retirement or permanent-cover project on their farm. This bid may have been equivalent to rent or opportunity cost (economist-speak for income given up for an alternative use).

Any additional costs such as trees, adjacent fencing, or a permanent grass cover could be added to a land rental equivalent or opportunity cost in their bid submission. This is where a producer could document in-kind, or a value they were willing to cover themselves to make their bid more attractive by keeping their request for funding lower. Keep in mind that the higher their bid per acre, the less likely they were to succeed, somewhat like a reverse auction. The only stipulation was that their bid could not exceed $10,000. The formula given to OSCIA committees would prioritize the bids, starting with the highest land-retirement priority in their county/district, but at the least cost. Their mandate was simply to find the best value for the funds requested.

Few producers received the maximum cost-share of $10,000. Often, the highest-quality projects that provided the biggest bang per buck didn't actually require that much funding. This formula factored in a landowner's contribution and on some projects, this was significant. Producers could receive funds for retiring and protecting fragile land, especially farmland adjacent to streams and open ditches. Called buffers, these vegetated strips of land act as a filter to catch silt and nutrients from adjacent crop land under agricultural production. Long-term agreements of five, ten, or fifteen years were signed with recipients; over ninety per cent of these were for fifteen years.[40]

Eligible projects included:

- Eight- to twenty-foot buffer strips with permanent grass and/or trees;
- Enhanced buffers which could include fencing to keep livestock from the watercourse;
- Block plantings of trees of up to twenty acres on highly erodible and/or fragile land; and
- Flood plains retired from crop cultivation.

Local OSCIA committees examined each applicant's proposal (bid) with the applicant's name and address hidden. When the approved list was calculated and the names made known to the committees, many committees spent a day or two touring their county/district to inspect the proposed projects to ensure the paper bid corresponded to the landscape. Nearly forty per cent of the projects were denied as these fell below their local allocated funding threshold.

The NSCP (Ontario) was developed to fill in the gaps in sustainable land management that the 1987 Land Stewardship Program didn't address. It provided funding for buffer strips along water courses, tree plantings on fragile land, and retirement of flood plains, all areas vulnerable to soil at risk.

---

40   National Soil Conservation Program Final Report. http://agrienvarchive.ca/nscp/download/nscp_final_rep.pdf, p. 3 (Accessed April 21, 2017). There were three parts to this section of the program:
   I.   Education through demonstration sites
        Original allocation: $605,000
   II. Administration and Communication
        Original allocation: $2.06 million
   III. Financial incentives for permanent cover
        Original allocation: $5.85 million, divided among the counties on the basis of row crop acres.

NSCP by the Numbers

The NSCP supported 1,226 projects covering 5,016 acres at an average cost of $979/acre, for a total expenditure of $4.9 million. Although the landowner agreements with OSCIA were not binding beyond fifteen years, it was anticipated the projects would become permanent features on the landscape. Here's what was achieved:

- Assuming a five-metre (sixteen-foot) width, over 1,600 kilometres (1,000 miles) of buffer strips were established in Ontario;
- 1,226 bids were approved out of more than 2,000 submitted;
- Flood plain retirement accounted for seventy-four projects on 231 hectares (570 acres);
- 2.5 million trees were planted on fragile farmland, and
- Forty-two kilometres (twenty-six miles) of windbreaks were established.[41]

A follow-up audit ten years later verified that virtually all the projects were intact and doing well with the intended purpose of removing fragile agricultural land from crop production.

2.5 Million Trees

How did we assemble sufficient quantities of the right tree species to match the soil, drainage, and climate conditions for a three-year program when it takes three to four years to establish tree nursery stock? The Ontario Ministry of Natural Resources (MNR) became a significant partner to make it work.

MNR agreed to provide the following services:

- Site visits to determine the tree requirements;
- Planting suitable nursery stock;

---

41 Ibid. p. 3 (Accessed April 21, 2017)

- Site preparation according to best practices one year ahead of planting;
- Planting trees when they became available;
- Controlling vegetative competition; and
- Assessments of plantings to ensure establishment of the trees.[42]

> This administrative arrangement got trees in the ground to match planting stock supplies with site conditions, flowed the federal funds on time and on budget and everyone was happy

For the entire process to work smoothly from preparing the land to building tree nursery capacity to planting the trees, landowners would sign a Tree Service Contract with OSCIA for the MNR's services. The landowner would pre-pay for these services and this proof of payment was sufficient for OSCIA to compensate the producer for the MNR follow-up. Some of these expert services, such as maintenance, replanting, or trimming, would be carried out several years later. OSCIA signed a Memorandum of Understanding with MNR and held the funds in trust until the services were complete. This administrative arrangement got trees in the ground to match planting stock supplies with site conditions, flowed the federal funds on time and on budget, and everyone was happy. It was a win-win service to the farm community and our government partners.

Formula for Success

OSCIA's management of the communication, administration, and delivery of farmer incentives under the permanent cover programs was a great success. It demonstrated that producers required substantially more than the previously offered $50 per acre. Further, as under the provincial LSP, when producers were empowered to organize their local program, they came forward with creative and cost-effective ideas for retiring marginal or fragile lands. I recall a few high-value per acre land retirement projects on vegetable

---

42   Ibid. p. 4 (Accessed Feb. 10, 2017)

land where opportunity costs were high, but the local committees also had sufficient checks and balances in place to justify those approvals. Conversely, there were exceptionally generous offers by landowners who requested no opportunity cost and contributed their own time and resources to land retirement from agriculture. These low-cost solutions were easily justified for approval and stretched the available funds over more acres.

The Green Plan

The Canada-Ontario Green Plan helped OSCIA build on the successes of the LSP and permanent cover programs by providing over $30 million for programs in Ontario. The five-year program, which ran from 1992 to 1997 and was delivered by AAFC and OMAFRA, had three broad objectives:

1. To conserve and enhance the natural resources that agriculture uses and shares;

2. To be compatible with other environmental resources that are affected by agriculture; and

3. To be proactive in protecting the agri-food sector from the environmental impacts caused by other sectors and factors external to agriculture.

The Green Plan consisted of sub-programs covering the following themes:

- Best management practices
- Rural conservation clubs
- Environmental farm plans
- Wetlands/woodlots/wildlife
- Research and technology transfer[43]

---

43   Bowman, B.T., M. Kingston. H. Rudy, C. Bradley. "Canada-Ontario Green Plan" http://agrienvarchive.ca/gp/download/gpsum97b.pdf (Accessed March 27, 2017).

The Retirement Years: National Soil Conservation and Permanent Cover Programs

The Green plan was significant to OSCIA in that it maintained the momentum of conservation programs established by the LSP and permanent cover programs. Further, it provided continuity for over six experienced Guelph office staff who had become proficient in the many aspects of program delivery. Additionally, the fifty-three county/district peer committees of farmers and field staff were well-established in grassroots program delivery across Ontario.

In particular, the Environmental Farm Plan (EFP) was one of the key success stories to emerge before the Green Plan ended in 1997. The EFP:

- Enrolled more than 10,000 participants;

- Continues to help farmers adopt more sustainable practices under the Growing Forward framework, with 35,000 participants to date;

- Strengthened partnerships between producers, government agencies, and agri-business;

- Provided guidance to farmers in more than twenty-five countries; and

- Helped unite the agriculture industry in communicating with government and the public. [44]

The EFP has provided land managers with a comprehensive review of all environmental concerns. It also brought together an exemplary collaboration of a private/public partnership. A full discussion is dedicated to EFP and its many achievements in Chapter 6. Before we delve into the broader scope of EFP, the soil management revolution that began in the 1980s requires a more in-depth review, leading up to the comprehensive work of modern-day soil fixers.

---

44   Bowman, B.T., M. Kingston. H. Rudy, C. Bradley. "Canada-Ontario Green Plan" http://agrienvarchive.ca/gp/download/gpsum97b.pdf (Accessed March 27, 2017).

Chapter 4

WHO GOT HIGH ON THE HIGH CROP RESIDUE PROGRAM?

"When government decided to pay $25 per acre for residue soil management, that was a big deal and got farmers really thinking about reduced tillage and conservation. We have OSCIA to thank for that."

-Lloyd Crowe, 2003 OSCIA President (Prince Edward County)

Who did get high on the High Crop Residue Program? Soil conservationists, that's who! In 1992, Agriculture and Agri-Food Canada (AAFC) approached OSCIA, indicating they had unallocated funds from the Land Management Assistance Program (LMAP). Could OSCIA design a program to encourage farmers to retain more plant residue (leaves and stems) from the harvested crop on the soil surface? By this time, the soil fixers had proven that plant residue, living or dead, would reduce soil erosion. They understood that when rainfall or melting snow runs across the fields, leaves and plant stems on the soil surface work like little filters or dams to slow the water down and reduce the water's erosive forces. The maximum benefit is achieved with no-till management: farmers don't plow or bury the plant residue—it's left on the soil surface. An added advantage is that after several years of no-till, earthworms become prolific, increasing the worm holes or porosity to absorb more rainfall rather than have it gush across the surface. Again, like in so many other conservation practices, soil protection from the erosive forces of water is an integral part of responsible soil management.

The no-till movement was just catching on after the introduction of the Land Stewardship Program (LSP), so here was an opportunity to reward farmers for the extra expense and risk of transitioning to no-till farming. The motives were solid. Pay farmers for results, not just incentives to buy the equipment up front. As you will see in this chapter, there are many dedicated people researching and adopting a multitude of practices to fix our soil, both on the surface and what lies beneath.

OSCIA staff and directors, along with our soil conservation adviser colleagues at OMAFRA, thought long and hard about an efficient way to measure residue and provide incentives on a per-acre basis. The process had to be fair, transparent, and defendable. The program could also be a great awareness opportunity for training farmers to recognize the level of soil protection provided through plant residue management.

Ontario's brilliant soil conservation advisers picked up on a concept developed by the U.S. Natural Resource Conservation Service as a simple way to measure the plant residue. Here's how it works: Take a twenty-foot-long rope and tie fifty knots spaced evenly apart, roughly every five inches. Peg one end into the ground and stretch the rope diagonally across the crop rows. Count the knots that touch plant residue and multiply by two. The result indicates the percentage of the soil surface covered by plant residue. Under no-till, 80 to 90 per cent residue can be achieved. Using other conservation tillage tools, 30 to 40 per cent residue was more realistic. Of course, the previous crop grown was also key to the amount of plant stems and leaves.

The High Crop Residue Program provided the following assistance to producers:

- $25 per acre for eligible acres where 30 to 39 per cent of the soil surface is covered by residue from previous crops at time of planting. $30 per acre was paid for eligible acres with 40 per cent or greater coverage;

- Up to 30 per cent of an applicant's total planted acres in the previous crop year was eligible, to a maximum of one hundred acres per program year; and

- The maximum contribution per applicant was $3,000 per program year ($6,000 over the two-year program).[45]

OSCIA equipped about fifty field staff across Ontario with residue management kits. I still have one of the original kits from 1993 displayed on a shelf in my office. Farmers were most intrigued to see field staff hovering over a knotted rope, counting the knots which intersected with plant residue. If they met the residue percentage requirement, the farmer was happy to hear that a cheque would be in the mail. Tremendous awareness and understanding of soil management occurred through this program.[46]

> Farmers were most intrigued to see field staff hovering over a knotted rope, counting the knots which intersected with plant residue

Figure 4.1 Knotted rope superimposed over 2018 no-till corn field showing 65 per cent plant residue protecting soil (Photo source: Christine Brown, OMAFRA)

---

45 Land Management Assistance Program. http://agrienvarchive.ca/lmap/lmapprog.html#High_Crop_Residue (Accessed March 27, 2017).

46 Ibid. A total of 3,638 out of 4,348 applications were accepted for funding. Field staff inspected 158,650 acres for crop residue. Approximately $4.8 million was paid out over the life of the program with an average payment per applicant of $1,462.17.

No-till management has evolved since the days of the original High Crop Residue Program to the point where its practitioners aim to leave the maximum amount of plant residue for soil protection. OSCIA farmers report that no-till works great where soil has excellent drainage and is particularly well-suited to winter wheat and soybean plantings. Close to 80 per cent of Ontario soybean and wheat crops are now planted using either no-till or other forms of conservation tillage, according to Dr. Bill Deen, a cropping systems research professor at the University of Guelph.

Technological advances in the twenty-first century threaten future use of the knotted rope where it may soon hang on a hook as an artifact in our museums. A millennial named Amelia, who is advancing her career in Silicon Valley, Calif., suggests that developing a plant residue-measuring app for her smart phone is no problem. With little time commitment and a quick digital scan aimed at the soil cover while walking across a farmer's field, the app will provide the correct percentage reading of plant residue. Amelia has been profoundly inspired by her mother, Christine Brown, an OMAFRA manure management specialist and long-standing conservationist working out of the Woodstock office.

Corn, however, continues to be a challenge on some farms. Heavier clay soil with imperfect drainage, particularly in the cooler regions of Ontario, often results in less than ideal crop emergence and lower yields. Good corn crops require warm soils at planting time for quick germination in early spring. Plant residue on the soil surface can leave the seedbed cold and wet. To be successful, no-till requires soil that is well-drained, either naturally or from engineered sub-surface drainage.

OSCIA members have carried out many on-farm field experiments over the years to test new and better corn planting methods that leave most of the stalks on the surface for soil conservation. One of the greatest challenges with no-till planting is to maintain yields of crops similar to traditional methods that involve fall plowing. Soil covered by straw and plant debris does not warm as quickly in the spring. This is critical for corn, which requires warm soil as early as possible for uniform germination and quick emergence. To overcome this challenge, scientists have shown that using a series of coulters (attachments for seeding equipment) to cut a band of strips across the field in the fall or early spring creates shallow ridges of bare soil that allow the seedbed to warm up quickly in the spring. This method also leaves lots of plant residue on the surface adjacent to the strips where seeds are slotted into the warmer soil. Using precision agriculture tools such as auto-steer, the corn planter seed openers follow the ridges precisely. The warm (ridged) seed bed provides equivalent yields to the old-fashioned way of plowing every inch of soil in the fall, which left the soil vulnerable to erosion.

> Indeed, there have been extraterrestrial impacts on our Ontario soils

Of course, where soil erodes, nutrients get transported from the fields into the waterways. Displaced nutrients result in less productive soil and polluted rivers and streams. We have made great progress in the past thirty years, but there are still barriers to overcome. Most farmers understand that soil care is a system that includes moisture management, crop rotation, a weed management strategy, and precise operation of equipment with suitable coulter configuration. Additionally, feeding crops with nutrients of the right amount, at the right rate, the right time and from the right source, known as the 4R Nutrient Stewardship Program, is also key.[47] Breeding plants for genetic resistance to damage from insects and disease is also critical. How will climate change affect these pests and what role will future genetic editing play? It's a system where all ingredients fit together for a complete recipe.

The Cover Crop Revolution- Wanted: Residue Dead or Alive

Planting of a green cover crop (often known as green manure) to protect the soil after harvest is not a new concept. For years, it has been done following vegetable and tobacco harvests. Over the past few years, there has been an extensive renewal in commitment to cover crops. OSCIA, with its member network and regional associations, is at the leading edge. Farmers are experimenting with many combinations of cover crop species.

---

47   Fertilizer Canada. "4R Nutrient Stewardship Program." https://fertilizercanada.ca/nutrient-stewardship/ (Accessed April 25, 2017).

What is a cover crop? OMAFRA explains: "Cover crops can do exactly what their name implies: cover the soil. A cover crop such as rye is commonly used to cover and protect the soil surface from wind and water erosion. The top growth covers the soil surface while the roots bind and stabilize the soil particles."[48]

> Alfalfa is considered one of the greatest soil builders and protectors from erosion

Cover crops are generally planted after a commercial crop is harvested, usually mid-summer or early fall, and grow up until freezing. Research continues to determine the best species or mix of cover crops. OSCIA members are experimenting with many choices and combinations of mixed cover crop species. Multiple species may be more beneficial than one species by itself.

Here are a few examples from OMAFRA's website:

- "Oilseed Radish: A great scavenger of leftover nitrogen from the commercial crop to convert the nutrient into organic matter for the next year's crop, preventing the leftover nitrogen from contaminating the surface or ground water.

- Ryegrass: Also a scavenger of nitrogen, it releases the leftover nitrogen later in the following year's season, compared to Oilseed Radish which releases the nitrogen earlier in the spring.

- Legumes: Clover, peas, alfalfa, etc. capture nitrogen from the air to make it available for next year's crop. The advantage of deep-rooted legumes such as alfalfa is that they go deep to salvage nutrients, break up soil compaction, and they're an excellent source of protein for livestock. Alfalfa generally requires a year of establishment and is well-suited as a commercial crop.

- Habitat for beneficial insects and insect control: Where sufficient growth time allows a cover crop species to flower, such as canola, red clover, sunflower, etc., the habitat may attract pollinators and/

---

48   OMAFRA. "Cover Crops: Adaptation and Use of Cover Crops." http://www.omafra.gov.on.ca/english/crops/facts/cover_crops01/cover.htm (Accessed April 25, 2017).

or predatory insects to reduce populations of damaging insects. Other plants, such as mustard, have fumigant properties where they can be grown weed-free for a full season. There is much research yet to do to better understand cover crops and in 2016, there were a record number of on-farm trials with various cover crop species, mixes, inter-cropping, and management practices."[49]

Extraterrestrial Soil Influences

Indeed, there have been extraterrestrial impacts on our Ontario soils. I'm not referring to biodynamic farming theories where cosmic forces are believed to be at play, although I've met a few farmers who are followers of this philosophy. Instead, I'm referring to the dramatic hit by a comet some 1.85 billion years ago that created a swath of fertile farmland within an impact crater sixty kilometres long and twenty-seven kilometres wide now known as the Sudbury Basin.[50]

I've visited farms in the Sudbury Basin several times during my career. One cannot help but be amazed at its sandy soils and flat, productive farmland, surrounded by the rocky, rugged rim around the edge of the crater, which is also known for its rich mineral deposits. It is the potato and sod farms in the basin that I find most fascinating.

Among the most productive farms in the basin, one is owned by Don Poulin Potatoes Inc., a fifth-generation family farm near Azilda, Ont. A delegation of OSCIA directors, staff, and family members toured the operation in August 2016, hosted by Dan Poulin and his sister Louise Mullally. They described in some detail how they manage about 1,000 acres at various locations in the basin. Each year they plant about 400 acres of high-quality potatoes, while leaving the rest of their land in hay or planted with cover crops. Depending on soil conditions, the Poulins' system calls for a second year or

---

49   Ibid.

50   Laurentian University. Goodman School of Mines. "Goodman Challenge." https://laurentian.ca/goodmanschoolofmines/ggc-about/ (Accessed Aug. 1, 2017) Until recently, scientists thought the crater was caused by an asteroid.

even three years of cover between potato crops. Rye-grass is preferred because it produces a small hay crop. Dan Poulin reports that their strong focus on soil health and soil management provides a 20 to 25 per cent increase in marketable potato yield, a lot less disease (meaning less use of pesticides), and more resilience to drought. This OSCIA member has the cover crop recipe figured out! Dan's late father Don was active on the OSCIA land stewardship committee and I had met him numerous times over the years. The Poulins are well-known in the community and receive lots of publicity in the local media for their great potato quality and respect for the soil.[51]

> The carbon cycle is essential to keep our planet habitable, and agriculture has an essential role to play in reducing overall greenhouse gas emissions through farm practices that sequester carbon, enrich the soil, and contribute to cleaner air and water

Potato farmers may be in the minority of OSCIA members; however, for those who have livestock, hay crops such as alfalfa serve the same purpose as ryegrass on a vegetable farm. Alfalfa is considered one of the greatest soil builders and protectors from erosion.[52]

The soil conservation ethic seen on the Poulin farm is typical of many of OSCIA's family farms. We can't rely on the next big comet or asteroid to create more farmland out of rock — the side-effects of such an impact would be catastrophic and may require a few hundred years (or perhaps a few million) for the air to become breathable again. So all hands are on deck to protect and build on the quality of soil that we have—a more predictable and gentle approach.

---

51 South Side Story. "Don Poulin Potatoes Family Business is the 'Pride of Azilda' https://southsidestory.ca/2016/11/don-poulin-potatoes-family-business-is-the-pride-of-azilda/
(Accessed April 25, 2017).

52 OMAFRA. Fact Sheet: "Cover Crops: Alfalfa" http://www.omafra.gov.on.ca/english/crops/facts/cover_crops01/alfalfa.htm
(Accessed Aug. 2, 2017).

Carbon Sequestration: Another Near-Extraterrestrial Influence

Since 2001, I have been involved in various administration duties working with extension staff and researchers to learn how soil can sequester more carbon and reduce concentrations of greenhouse gas (GHG) emitted by our society. Carbon is sequestered in the soil from plants taking in carbon dioxide (CO_2) through photosynthesis. Nitrous oxide and methane are the other two greenhouse gases in the atmosphere of concern to agriculture. Both cause damage to the ozone layer of the earth's atmosphere that protects living organisms from harmful solar radiation and helps regulate the weather.[53] OSCIA's focus has been primarily on CO_2 with its connection to carbon in the organic matter of soil.

In short, most everything contains carbon and farmers want to keep it in the soil as healthy, rich, black organic matter rather than have it escape into the atmosphere in the form of carbon dioxide. Too much tillage destroys the soil carbon and contributes to CO_2 in the atmosphere.

Conversely, where carbon can be captured and stored (sequestration) via more plant growth, less soil movement, and the accumulation of biomass within the soil, everyone wins. The carbon cycle is essential to keep our planet habitable, and agriculture has a critical role to play in reducing overall greenhouse gas emissions through farm practices that sequester carbon, enrich the soil, and contribute to cleaner air and water.

In 2001, OSCIA and the Soil Conservation Council of Canada (SCCC) launched the Greenhouse Gas Mitigation Program (GHGMP). Other regions of Canada also aligned with SCCC to focus on their regional opportunities in agriculture.

In Ontario, we established five demonstration farms as hubs, with additional farms nearby focusing on similar field work. In total, our project in Ontario lined up over fourteen demonstration farms from the Ottawa Valley in the east to Essex/Kent counties in the southwest. These farms demonstrated Best Management Practices (BMPs) with potential to reduce GHG.

---

53   Earth Topics, NASA, https://www.nasa.gov/topics/earth/index.html (Accessed May 5, 2018)

A number of projects focused on 'Nitrogen Use Efficiency' in horticultural crops. Others looked at precision no-till planting on heavier clay soils where no-till was difficult for farmers to adopt. Still other demonstration farms looked at various species of cover crops to measure the increase in biomass (hence increase in organic matter) that could be gained from application of manure to stimulate growth in cover crops.

For many producers, the on-farm research through the GHGMP was their first look at the practical ways farms can contribute to fighting climate change. One example illustrated how 40–60 kilograms of manure nitrogen per hectare was sequestered into biomass with the use of cover crops after harvest, and where the nitrogen could be saved for next year's crop. For others seeking to adopt residue management on the challenging heavy clay soils, a new form of precision planting into previously prepared strips could result in yields similar to those of conventional plowing and tillage practices in corn.[54]

While the formal GHGMP program ended around 2007, the work to develop best practices for GHG mitigation has continued. Research is long term and we are continually seeking ideas for continuous improvement.

The GHG issue is of such critical importance that the Food and Agriculture Organization of the United Nations (FAO) organized the Global Symposium on Soil Organic Carbon, held in Rome in March 2017. The symposium featured 103 scientific presentations and thirty-five poster presentations on measuring, managing, and storing carbon for the health of our soils and our planet.[55]

Strong research support continues in Ontario today, where OSCIA members are taking part in on-farm investigations led by OMAFRA specialists under the Best Management Practices Verification and Demonstration Fund. Current examples include:

---

54   Greenhouse Gas Mitigation Program Report, OSCIA internal archives

55   FAO. "Proceedings of the Global Symposium on Soil Organic Carbon 2017."
http://www.fao.org/documents/card/en/c/d6555d8d-1b19-4c04-a25d-74474e6c0a11/
(Accessed Sept. 14, 2017).

- Christine Brown, OMAFRA manure management specialist, is examining the effects of soil amendments from methane digesters and other processed biosolids. These waste by-products are being applied to corn and post-harvest wheat fields, adhering to recommended nutrient rates. In the harvested wheat stubble, green cover crops are managed to increase maximum plant growth until winter. The research will measure, among other scientific parameters, the change in soil organic matter. It's hoped that these methods will boost soil carbon fixation rates (and in the process, reduce levels of carbon dioxide, our most common greenhouse gas).

- Adam Hayes, OMAFRA soil management specialist, is studying soil remediation with a focus on determining how to increase soil organic matter with manure or compost. The project explores how much compost/manure would be required to increase concentrations of organic matter in soil by one per cent. Preliminary calculations indicate that 300–400 tons per acre of leaf/yard compost, mushroom compost, solid dairy manure, or solid horse manure is required. That's equivalent to about 30–40 semi-trailer truckloads per acre! That much would overload fields with nutrients and contaminate the soil and nearby water supplies. Ideal amounts range from about one semi-trailer truckload of mushroom compost to three truckloads for dairy manure to achieve the proper balance of nutrients. These amounts would increase the organic matter by about 0.1 per cent.

There are no quick fixes to sequestering carbon and increasing organic matter. By adopting best practices using hay, pasture, and cereals in rotation with corn and soybeans, and utilizing no-till planting where possible, farmers can keep the microorganisms in the soil happy, which will in turn stabilize and increase the amount of organic matter over time. Adding soil amendments as described in the examples above will accelerate the rate of carbon sequestration.

Leading-Edge Research on Soil Critters to Understand and Improve Soil

There is exciting new research underway on the role of cover crops to improve soil health. These investigations aim to learn more about what may be happening within the microbe kingdom below the soil surface. In the past, we just haven't understood their complicated world. If we can make these billions of microbes feel at home, welcome, and motivate them to repair soil in and around the root zone, we will go a long way in solving healthy soil challenges.

> If microbes are our friends and you don't have any friends in your fields, how about looking in your fence rows or woodlots?

Recent research findings provide positive and optimistic strategies for soil fixers. Remember my earlier reference to billions of microbes in a teaspoon of soil? Through modern scientific methods, we are learning how to identify and measure these important soil microbes. Advances in gene mapping lead the way. These scientific discoveries are opening a new window on the universe below the soil surface that may hold new keys to sustainable soil management and food security.

For example, University of Saskatchewan's Dr. Kate Congreves focuses on the importance of soil biogeochemical pathways in influencing the formation of organic matter. In a presentation titled "You Want Soil Organic Matter!," Congreves told the 2017 South-West Agricultural Conference in Ridgetown, Ont., that microbial processing of crop residues is an important factor in forming soil organic matter.[56]

Dr. Bill Deen of the University of Guelph has been engaged with OSCIA on long-term tillage and crop rotation trials. He has authored numerous research papers and spoken at various functions in Ontario to communicate findings on the importance of crop rotations to improve soil quality. In particular, Dr. Deen manages a thirty-one-year, long-term tillage and crop

---

56   Soil Health, Biogeochemistry, Crop Nutrition, Kate Congreves, http://researchers.usask.ca/kate-congreves/ , accessed January 12, 2018.

rotation trial, which found that the most resilient soils — those that can withstand extremes in weather such as drought or excessive moisture — were those managed with a diversified crop rotation that included not just corn and soybeans, but also cereals under-seeded to red clover or alfalfa. In hot, dry years, when cereals/legumes were grown using reduced tillage (conservation tillage), the yield increased seven per cent for corn and 22 per cent for soybeans.[57]

Important living organisms contributing to soil health are listed in Figure 4.2 below, created by soil experts from AAFC and OMAFRA who presented their research at the FarmSmart 2017 Conference in Guelph, Ont. In a presentation titled "Soils Are Alive! When Biology and Agronomy Meet," AAFC's Dr. Lori Phillips and Jake Munroe of OMAFRA stated microbes:

- Cycle carbon through photosynthesis and decomposition;
- Regulate the supply and loss of plant nutrients including nitrogen, phosphorous, potassium and iron through symbiotic and asymbiotic reactions;
- Capture and release greenhouse gases (carbon dioxide, methane, nitrous oxide);
- Improve soil structure (aggregate stability);
- Degrade pesticides;
- Regulate water quality, for example by filtering nutrients; and
- Suppress soil-borne diseases.[58]

---

57   Gaudin, A.C.M. et al. "Increasing Crop Diversity Mitigates Weather Variations and Improves Yield Stability." PLoS One (Feb. 6, 2015) http://journals.plos.org/plosone/article?id=10.1371/journal.pone.0113261

58   Philips, Lori A. and Jake Munroe. "Soils Are Alive! When Biology and Agronomy Meet." https://farmsmartconference.files.wordpress.com/2017/02/17fs48-munroe-and-phillips-soils-are-alive.pdf. (Accessed April 25, 2017).

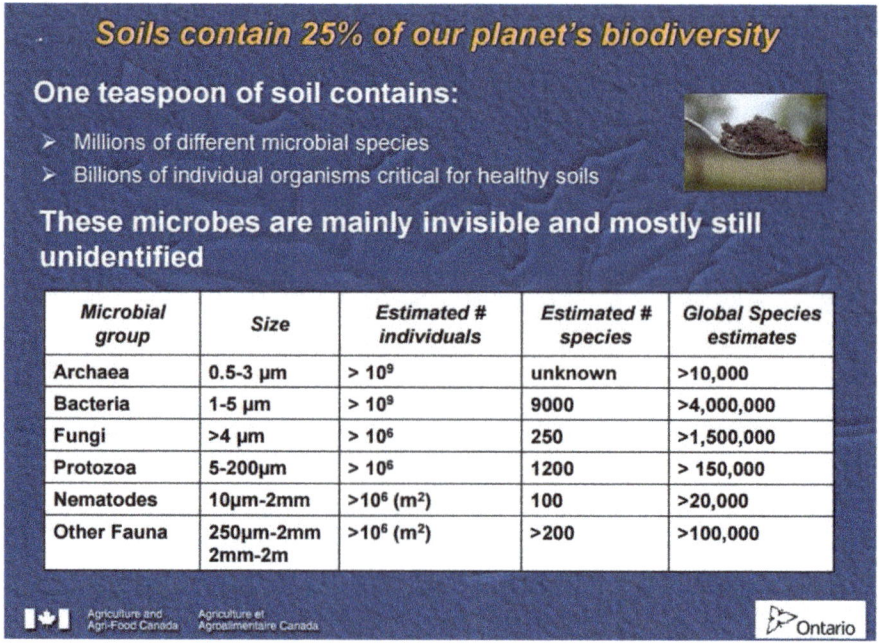

Figure 4.2 - Billions of organisms contribute to healthy soil.
(Source: Agriculture and Agri-Food Canada and the Ontario Ministry of Agriculture, Food and Rural Affairs, presentation at Farm Smart Conference, January 2017)

There you have it! Many microbes are our friends and we need to learn much more about them. Through ongoing scientific research and a variety of outreach and educational efforts, we are doing just that.

Fence Row Farming

If microbes are our friends and you don't have any friends in your fields, how about looking in your fence rows or woodlots? That's what happened on the farm of Dean Glenney over twenty years ago. In fact, he became known as "the fence row farmer," but over time, his whole farm demonstrated the same healthy soil makeup that he noticed in the fence rows. With the help of global positioning system (GPS) technology, he used the same equipment tracks year after year where the rows of corn and soybeans were not compacted from wheel tracks. The soil where the seeds were planted and the newly emerged plants and roots thrived with minimally disturbed soils. Mr. Glenney now has so many microbial friends that he was named the 2015 OSCIA Soil Champion[59] for his contributions as a field crop producer in the Dunnville area of Haldimand County.

Investing in Soil Management

Funding agencies continue to invest in scientists working to get a grasp on how to improve the microbial activities, monitor results, and easily measure soil health. Some of this work is led by Dr. Kari Dunfield, an associate professor of applied soil ecology at the University of Guelph who holds a Tier 2 Canada Research Chair in Environmental Microbiology of Agroecosystems. Her research is focused on understanding the environmental impacts of farming systems by using DNA analysis to study living organisms in soils and water and how key groups of microbes affect the functioning of soil ecosystems.

Ontario Soil Network

The Rural Ontario Institute has been leading the charge to focus on soil health with the launch of the Ontario Soil Network in March 2017. OSCIA joined other sponsors in supporting this initiative. This ten-month leadership and communications course brought together farmers from across southwestern Ontario who had implemented best practices on their farms. The goal is to train, inspire, and support innovators who are taking the lead in promoting soil health through no-till systems, planting cover crops, and more. Four of the fifty participants in the inaugural class are familiar names to OSCIA: Gord Green (2016 president), Alan Kruszel (2015 president),

---

59 OSCIA created the Soil Champion Award in 2014 to recognize those individuals who demonstrate exemplary commitment to soil conservation. Learn more about Dean Glenney and other winners in Chapter 11, Modus Operandi, or visit www.ontariosoilcrop.org/association/association-soil-champion-award/.

Warren Schneckenburger (provincial director, Eastern Valley Region),[60] and Steve Sickle (provincial director, Golden Horseshoe Region). The project continues with a variety of learning opportunities throughout the year.[61]

Airing Our Dirty Laundry

It must have been a long, cold winter a few years ago for the person who came up with this newly inspired method for measuring soil health. Was it Nebraska, the Dakotas, or South Carolina? No one is admitting to it! In any case, soil conservationists hypothesized that cotton cloth would be chewed up in record time when buried in soil that was blessed with abundant microbes. Conversely, if the numbers of microbes were low (poor soil health), cotton cloth buried in the soil for a couple months would remain intact. In the interests of good science, a comparison of several soil management histories could visually provide the contrast of great soil health (lots of microbes, cotton cloth eaten) versus poor soil health (few microbes available to eat the cotton cloth).

> One has to ponder the circumstances upon where and when the soil conservationist looked around and in desperation grabbed perhaps the only cotton cloth available—their underwear—and, as a result, a new soil analysis tool was born

One has to ponder the circumstances upon where and when the soil conservationist looked around and in desperation grabbed perhaps the only cotton cloth available—their underwear—and, as a result, a new soil analysis tool was born.

Anne Verhallen, OMAFRA soil management specialist, described to me how the cotton cloth test evolved in Ontario. She learned about the technique in early 2015 while attending a Midwest Cover Crop Council meeting in Iowa accompanied by University of Guelph Ridgetown Campus research

---

60   OSCIA counties/districts are combined into regions. Further discussion on regions can be found in Chapter 11, Modus Operandi.

61   Ontario Soil Network. "Let's Talk Soil" www.ontariosoil.net (Accessed April 25, 2017).

technician Claire Coombs and Woody Van Arkel, a soil sustainability advocate and farmer from Dresden, Ont. A presentation at the meeting featured a slide with badly frayed underwear that had been buried in the ground.[62]

After returning to Ontario, Anne headed to Walmart and bought a package of seven undies. Claire buried four in the college's long-term crop trials, one was used as a tester, and the other two were given to Blake Vince, a conservation farm leader from Merlin, Ont. Blake posted a video on YouTube of Claire burying the underwear. They were dug up several months later and displayed at an Innovative Farmers' Association of Ontario (IFAO) field day where they became a major attraction. IFAO is credited with coining the hashtag #soilyourundies, and the organization posted an instructional video on how to bury underwear for testing that inspired a contest at their 2016 winter conference. The underwear and photo submissions were displayed at the conference on a clothes line near the registration table. Twitter lit up and the call to "soil your undies" took off (pun intended) around the world.

Local OSCIA groups also got involved. The Renfrew county Soil and Crop Improvement Association initiated a soil health project by burying underwear to display at a later field day. Jean Sullivan's 4-H club used the underwear test in their display at the annual Carp Fair in 2016.

Alan Kruszel, Chair of the Soil Conservation Council of Canada (SCCC), brought Stanfield's on board as a sponsor and earned national media coverage when he launched Canada's Soil Your Undies campaign during National Soil Conservation Week in April 2017. The cameras captured the proper undie-burying technique as he challenged farmers across Canada to compare soil health with the comparative cotton test.[63]

In the summer of 2017, IFAO and Woody Van Arkel issued a challenge to farmers in North Dakota to see who could bury more underwear. Scientists,

---

62   Kent, Peter. "Dirt-y drawers: Researchers bury underwear to demonstrate unhealthy soil." The Newsstand. Clemson University (Aug. 20, 2014). http://newsstand.clemson.edu/mediarelations/dirt-y-drawers-researchers-bury-underwear-to-demonstrate-unhealthy-soil/ (Accessed Aug. 9, 2017).

63   Soil Conservation Council of Canada news release. http://www.soilcc.ca/news_releases/2017/nr_2017-03_e.php (Accessed June 30, 2017).

too, have become involved, with probing through the microbiology in the "undiesphere." The altered laundry will no doubt consume much time on the agendas of winter meetings in years to come.

Anne Verhallen reports that the original Walmart purchase of gotchies has travelled across Ontario with social media capturing the moments for viewers around the world. Anne says her only regret was that in the excitement of launching the new exploration, the original Walmart receipt was lost so she hasn't been able to claim the cost of "supplies" on her expense account.

Perhaps there is a business opportunity for farm equipment dealers to stock their shelves with cotton boxer briefs? On our farm however, we are a bit more modest and elected instead to bury several cotton handkerchiefs. I'm sure the microbes won't mind as they meet the 100 per cent cotton requirement for the experiment. Here we have yet one more tool for soil fixing!

Today, social media contains a plethora of soil fixers who openly share their experiences with no-till, cover crop options, crop rotations, and other farming details, often accompanied by a photo. These soil health advocates generously share their time and experiences for the good of our planet.

Chapter 5

Who Sets the Agenda for Agriculture?

When it came to environmental policy for agriculture through the 1980s, a number of things needed fixing. Who set the rules for how farmers grow crops and manage their poultry and livestock? Growing up on a Mennonite family farm in the 1950s and 1960s, my nine siblings and I did what we learned from our parents, as many other farmers did from theirs. Time, energy, and social commitment centered around the church and family, all of whom were mostly farmers. Our parents didn't attend field days, crop tours, or agricultural workshops. With ten kids, they likely didn't have time!

In hindsight, we missed lots of other learning moments as kids. 4H programs have been a wonderful training opportunity for rural children and teenagers to network and learn from each other. Commodity groups also had their workshops, tours, and conferences. This broader networking has proven valuable for adoption of advanced soil, cropping, and livestock management methods. In addition, this interacting developed leadership at the grass-roots often influential in establishing policy.

OSCIA had been working with farmers since 1939 and was connected to researchers at the Ontario Agricultural College (OAC) and government extension staff. This collaboration helped many farmers learn how to increase yields with better grain varieties and improved soil management. If an OSCIA member in the 1960s was lucky enough to travel to the OSCIA's annual meeting (usually held at the Royal York Hotel in Toronto), he or she would have heard advice from various speakers and farmers from across Ontario. This was a major conference during that era with often five hundred in attendance. It was well-accepted that OAC professors and government extension staff were the experts who provided valuable knowledge and expertise required for innovative farm management.

> *"I realize the important leadership OSCIA has provided through well attended field tours and annual meetings as well as outstanding conferences that provided leading-edge, practical ideas that farmers could take home and practice on their own farms to improve their operations while protecting the environment."*
>
> -John Benham, 1978 OSCIA
> President and EFP workshop leader (Wellington County)

Environmental concerns within agriculture with their roots in the 1980s began to be addressed in earnest by farm leaders in the 1990s. A coalition of farm leaders was growing concerned about the mixed messages being heard from non-farm advocates about agricultural practices. There were often conflicting views on how farming practices were impacting the natural environment.

How could farmers respond to these multiple agendas? This left farmers and farm organizations confused about the direction of future environmental policy for agriculture, or worse, worried that impractical and costly changes would be imposed upon them. Ontario's growing urban population was increasingly disconnected from agriculture, but also more and more concerned about environmental issues such as the impact of farming practices on water supplies.

In the 1990 provincial election campaign, the New Democratic Party (NDP) made numerous promises that they would take control of environmental issues if elected. The NDP platform included a proposed Environmental Bill of Rights, which with its vague criteria, was seen as a threat to modern farming practices. It is one thing to promise the world when politically campaigning; it is another to wake up and discover you're now the government in power, elected to deliver on those promises. Farm leaders were worried at the uncertainty when the NDP won a shocking majority, based in large part on its strong urban base.

Recognizing concerns in the agricultural community, Elmer Buchanan, the newly appointed minister of agriculture and food, set up a Ministerial Advisory Committee on Environmental Responsibility and invited farm leaders to attend. Numerous hearings were held representing many interests. Everyone appeared to have an environmental agenda for agriculture except, it seemed, the agricultural industry itself.

> Our Farm Environmental Agenda expressed the coalition's commitment to encouraging all farmers to conduct their activities in a manner that respects the environment

Farm leaders became extremely concerned and in response, a coalition representing more than thirty Ontario farm organizations held a series of meetings to determine a course of action. The group was led by Roger George, President of the Ontario Federation of Agriculture (OFA), Jeff Wilson, Chair of AGCare, Henry Aukema, President of the Christian Farmers Federation of Ontario, and Gord Coukell, Chair of the Ontario Farm Animal Council. Farm leaders suggested it was better to be proactive than reactive; to work together to resolve issues before they escalate into a crisis. It would not be constructive for farm leaders to respond after policy decisions or new regulations were thrust upon the industry. This coalition became known as the Ontario Farm Environmental Coalition (OFEC).[64]

It was suggested that someone with a neutral perspective, independent from the government and unconnected to any farm organization, was needed to chair the group. The coalition approached Dr. Gord Surgeoner, a University of Guelph entomologist who was becoming well-known for his practical perspective on agriculture and the environment. Dr. Surgeoner agreed to chair the steering committee and the working group, which included various staff

---

64   Kingston, M. et al. "Successful Partnerships in On-Farm Environmental Action in Ontario: The Environmental Farm Plan and Related Initiatives." p. 4. www.academia.edu/27847293/Successful_Partnerships_in_On-Farm_Environmental_Action_in_Ontario_The_Environmental_Farm_Plan_and_Related_Initiatives (Accessed April 25, 2017).

members of farm organizations, farm representatives, and government staff. Numerous sub-committees were also set up to tackle the tough issues.

This led to the development of a policy document called *Our Farm Environmental Agenda*. It expressed the coalition's commitment to encouraging all farmers to conduct their activities in a manner that respects the environment. It also outlined plans to ask farmers to develop and maintain environmental farm plans, beginning in 1992. Over the next decade, the agenda was expected to result in the creation of over 40,000 individual environmental farm plans, one for every farm business in the province.[65] Dr. Terry Daynard, then the executive vice-president of the Ontario Corn Producers' Association, was a major influence in bringing the agenda forward.

Farm leaders met Rita Burak, Ontario's deputy agriculture minister, to outline the industry's environmental policy proposal and request ministry support. Ms. Burak was fully supportive of the farm leaders' request. Ministry staff were committed to move the environmental farm plan concept forward in a collaborative process. Having consensus among farm leaders to speak as one voice to government was seen as a powerful tool in the political process.

The value of bringing together farm leaders and government in a collaborative process would continue to emerge in later years when other confrontational issues arose. Many municipalities were developing nutrient management requirements for building permits to strengthen standards governing approvals of new livestock or poultry barns. Although individual municipal by-laws were well-intended efforts to control manure management, the rules were not consistent across municipal boundaries. The result was a hodgepodge of inconsistent requirements and confused stakeholders. So in 1998, farm leaders from OFEC assembled a Nutrient Management Strategy working group to consult and collaborate with the province in developing consistent regulations under one piece of legislation, ultimately leading to the Ontario Nutrient Management Act. Regulations would prescribe building permit requirements for barns as well as new rules for manure storage and application on farms. OSCIA's role in delivering nutrient management

---

65 "Our Farm Environmental Agenda." p. 23. http://agrienvarchive.ca/gp/efp/agenda.html. (Accessed April 5, 2017).

cost-share programs to the farm community is discussed in detail in Chapter 9, "What Do Crops Eat for Breakfast?"

Similarly, when regulations were proposed prior to 2002 under the Safe Drinking Water Act (it became the Clean Water Act, 2006), farm leaders formed the Water Quality Working Group and took an active role in formulating a practical approach to implementation.[66] More recently, a coalition of agricultural groups, food and beverage manufacturers, government, and non-government agencies has begun to address issues of sustainability. Following a year of stakeholder consultations, the Sustainable Farm and Food Initiative (SFFI) was launched in 2017 to outline a path forward with a trusted, verifiable system of standards to meet growing consumer demands for farm and food sustainability.

The power of collaboration can't be overestimated when establishing government policy and implementation of regulations. OSCIA has been a beneficiary with its unique role in program delivery. The various collaborations and networking with stakeholders have opened the door to new programs and projects. It also helps OSCIA fulfill its mission to "facilitate responsible economic management of soil, water, air and crops through development and communication of innovative farming practices."[67]

This collaboration has been great for setting the agenda for agriculture and the environment. It has been a wonderful model for tackling other challenges and should form a case study for students in years to come. A major outcome has been the Environmental Farm Plan (EFP) program. In fact, the EFP has certainly been one of the highlights of my career and I'm thrilled to reflect on my involvement with it.

---

66   Clean Water Act (2006). www.ontario.ca/laws/statute/06c22 (Accessed April 25, 2017).

67   OSCIA. "About Us." www.ontariosoilcrop.org/association/association-about/ (Accessed Aug. 16, 2017).

Chapter 6
A WORLD-CLASS ENVIRONMENTAL FARM PLAN

"Being involved in the development and implementation of Ontario's EFP program was a highlight of my twenty-seven-year career with the Ontario Federation of Agriculture. The experience served as an amazing example of how general farm organizations, commodity-specific farm organizations, provincial ministries and federal departments were able to set aside any competing interests and work collectively to achieve an outcome of benefit to the agricultural sector, in particular, and all of society in general. The fact that the EFP is still relevant all these years later is a real testament to the work of all of those involved in the process."

- David Armitage, Director of Regulatory Reform,
Ontario Federation of Agriculture

Farm planning was a concept I learned about when I joined the provincial agriculture ministry in the mid-1980s. What is farm planning? How does it fit with all the activities on the farm and how can it be used for sustainable food production? Various perspectives were brought forward during meetings and coffee-shop talk.

Early discussions about farm plans revolved around preparing a map for each farm field identifying the soil type, crop rotation, soil testing for nutrient availability, drainage, and topography. Through use of formulas and equations, a prescription could be drawn up to outline how best to manage the crops to eliminate soil erosion. Preparing farm maps would involve a technical expert—an engineer, soil scientist, or trained technician—who would visit the farm, interview the owner, draw maps of existing topography including soil type and drainage features, identify crop rotation, document tillage, fertilizer or manure application to suit the crop, and planting equipment. The experts would present their findings and include recommendations in a report for the farmer. Recommendations might include erosion control structures such as grassed waterways or tile inlets, which required further design work involving setting up a transit to survey the fields. This was the early concept of a farm plan. Although quite time-consuming, it was thorough.

The most common formula to calculate potential soil erosion is called the Revised Universal Soil Loss Equation for Application in Canada (RUSLEFAC),[68] which is still in use today. Throughout the 1980s, several dozen Ontario farms may have been surveyed using an earlier version (called the Universal Soil Loss Equation) of the RUSLEFAC approach where various scenarios of crop rotation, tillage, etc. could be laid out in a prescription developed by the expert. Best practices for each field would identify what crops to grow, how to till the soil, which fertilizer to use and where to install tile drainage. Water management could also include erosion control structures, such as grassed waterways, surface water inlets to tiles, and tile outlet protection.

This type of planning was labour intensive, with expertise provided by agricultural engineers from the provincial government or local conservation authorities, supported by trained technicians. Participating farmers enjoyed this free service and were very pleased with the professional recommendations. The farm plan presented to the farmer consisted of a booklet with lots of photos and maps. Could staff resources be expanded to accommodate EFP for each farm, as promised by farm leaders in *Our Farm Agenda*? Were there alternative approaches to an EFP?

---

68 Wall, G.J. et al. "Revised Universal Soil Loss Equation for Application in Canada." Research Branch, Agriculture and Agri-Food Canada. www.sis.agr.gc.ca/cansis/publications/manuals/2002-92/rusle-can.pdf (Accessed June 21, 2017).

The message coming from government was that there would not be sufficient staff resources available to provide individual service to every farm. After all, there were close to 60,000 registered farms in Ontario in the mid-1980s. Other approaches were being tested. The Land Stewardship Program (LSP) was our first attempt to provide a broad-based farm plan as a prerequisite to cost-share funding. The LSP farm plan was nothing more than a few pages of questions to be answered by a producer. They would identify the need for a land stewardship cost-share grant, and identify the project location on the farm sketch. Typically, the LSP farm plan would take only a few minutes for the farmer to complete. Although somewhat limited in its scope, purpose and function, it was the start of a broad-based environmental farm planning initiative that relied on self-assessment. Who better to understand their problems and needs than those working the land? If farmers received training to clearly understand best agronomic practices, could they not complete the self-assessment and identify requirements for improvements?

Looking South of the Border

Committing to an environmental farm plan was challenging because nobody really knew what it should look like. What self-assessment could address all environmental issues? An engineer's report showing field maps and topographical features was a start, but it did not address many management practices such as pesticide storage and application, integrated pest management (IPM), drinking water safety, nutrient application, or petroleum storage.

Under the direction of the OFEC Steering Committee, I chaired a subcommittee to develop an EFP that met clear criteria. An EFP must be:

- Comprehensive: it had to address all environmental concerns related to farmsteads, livestock / poultry, field operations, and surrounding woodlots, streams and wildlife areas.
- Voluntary: no one would be forced to complete an EFP.
- Confidential: environmental concerns identified in the plan would not be passed on to enforcement agencies.

- Self-directed: each farmer was to complete his/her own assessment after an introduction to best practices blended with a thorough training program.

- Credible: adults learn differently than young children, therefore content had to be technically accurate and credible to ensure a positive learning experience.

- Understandable: all documents and materials had to be created using plain language.

- Performance-based: it must define agronomic and environmental best practices reached by consensus with multiple agencies and farming practitioners.

- Relevant: it must outline relevant regulations to help farmers understand whether they are in compliance.

- Practical: it must flag areas of concern and include an action plan with a timetable for improvement.

- Success-oriented: the plan must identify and document positive environmental achievements; there are many good news stories to be shared.[69]

Our committee struggled to envision an EFP that would meet all of the above requirements. There was an "Aha!" moment when the manager of OMAFRA's engineering unit, George Garland, introduced us to the Wisconsin Farm*A*Syst program, developed by the University of Wisconsin. This self-assessment system enabled farmers to rate their management practices in a number of environmental categories, such as risk to groundwater.

However, the Farm*A*Syst workbook initially focused only on reducing risks to groundwater. It did not cover field activities, nor the surrounding woodlots, wetlands, streams, ditches, or wildlife areas. With permission from the Wisconsin group, we recommended that the EFP would be based upon the Farm*A*Syst model but modified to ensure it was relevant to Ontario conditions. OMAFRA provided essential expertise to help draft the

---

69   OSCIA Archives.

workbook and ensure that each of its twenty-three sections was technically accurate. OMAFRA staff mostly chaired subcommittees of farmers and other experts who were responsible for each section. A list of these first twenty-three committees can be found in Appendix 11.

Brent Kennedy (now director of OMAFRA's Rural Programs Branch) headed up all these sub-committees and, with expertise from graphic artist David Berman and plain language professional Ruth Baldwin, produced an award-winning, world-class EFP workbook. In fact, Mr. Berman informs me that over the years, he has included the EFP as a sterling case study in his keynote talks at design and literacy conferences on five continents. Don Hill, OSCIA's 1988 President, was hired to co-ordinate the county/district training and promote the program at various events.

EFP Content and Delivery

The EFP Workbook was comprehensive, containing twenty-three sections that addressed key topics ranging from handling and storage of pesticides to nutrient management to milking centre wash water to woodlands and wildlife.[70]

Another important component of the workbook was the Action Plan. Where farmers did not comply with best practices as defined in the EFP Workbook, they would flag it as an area for improvement and identify what

---

70 Ontario Environmental Farm Plan Workbook (2017). As of 2017, the section titles are: Soil and Site Evaluation; Water Wells; Pesticide Handing and Storage; Fertilizer Handling and Storage; Storage of Petroleum Products; Disposal of Farm Wastes; Treatment of Household Wastewater; On-Farm Storage, Treatment & Management of Manure & Other Prescribed Materials; Disposal of Livestock Mortalities; Storage and Feeding of Ensilage; Milking Centre Washwater; Nuisances and Normal Farming Practices; Water Efficiency; Energy Efficiency; Soil Management; Nutrient Management in Growing Crops; Use and Management of Manure and Other Organic and/or Prescribed Materials; Horticultural Production; Field Crop Management; Pest Management; Stream, Ditch, and Floodplain Management; Wetlands and Wildlife Ponds; and Woodlands and Wildlife.

they intended to do to correct the deficiency in their Action Plan. A timeline for improvement was also required.

Having a world-class EFP workbook on the shelf was one thing, but getting it into the hands of farmers for their participation was another. Once again, OSCIA was assigned the task of marketing the program and engaging producers. County/district committees of farmers, their peers, were charged with local delivery.

The local OSCIA committees' work with the Land Stewardship Programs (LSP) and the National Soil Conservation Program (NSCP) provided a highly successful delivery model. Why not empower similar committees to follow the same model for EFP delivery? We called them peer review committees. One of the objectives was to build a quality-control mechanism into the EFP program that could challenge a farmer's self-assessment or suggest additional resources or ideas to make improvements. One of the most valuable traits of the local farmer delivery concept was that committee members took ownership of their duties and promoted the concept among fellow farmers and agri-business contacts in their communities.

> One of the most valuable traits of the local farmer delivery concept was that committee members took ownership of their duties and promoted the concept among fellow farmers and agri-business contacts in their communities

I credit our peer review committees for the early success of EFP. The first few years were tough sledding. There were naysayers in the farm community who said publicly that they would never do an EFP, which they viewed as tantamount to confessing their environmental sins on paper. Confidentiality was a big deal and their greatest fear was to have an inspector from the Ministry of Environment show up on a fishing exercise and demand their EFP to look for environmental offences to prosecute.

To roll out the initiative, local OSCIA and OMAF staff members led two-day workshops aimed at helping producers to understand environmental best practices and regulations. Farmer participants were assigned the task of conducting a thorough self-assessment of their operations and completing action plans that identified and addressed deficiencies or areas of high risk.

The self-assessments and action plans were checked for completeness by the workshop leaders, and then reviewed anonymously by the peer review committee for final endorsement. Completed action plans were required for cost-share funding to address concerns identified through the EFP process.

Although there was understandable reluctance to attend workshops in some circles, the EFP was a hit with those farmers attending right out of the gate. Participants gave it a big "thumbs up" in a survey that followed the first workshops held as part of the 1993 pilot project. Asked if they would recommend the EFP process to a neighbour, 95 per cent said yes.

> In the end, not only did Dr. Surgeoner receive a letter of assurance from the Ministry of the Environment and Energy (MOEE) that an EFP would not be used as evidence, the ministry also developed a document called Policy Guidelines on Access to Environmental Evaluations (November 1995)

The initiative was so successful that on several occasions, additional field staff were hired as program representatives to review the many EFP workbooks and action plans. They were submitted by the thousands during the peak years of 1997-2000 and by the turn of the new millennium, 20,000 farmers had voluntarily enrolled in EFP.

EFP was also earning kudos from several prominent organizations, including the:

- Recognition Award - SOLEC (State of Lakes Environmental Conference), 2000
- Pollution Prevention Leadership Award - Ontario Ministry of Environment and Energy, 1997
- Blue Ribbon Award - American Society of Agricultural Engineers, 1994.[71]

---

71   OSCIA Archives.

Confidentiality

Farm leaders on the steering committee were also worried about confidentiality, so much so that Dr. Gord Surgeoner went straight to the attorney general's office in Toronto to lay the cards on the table. He said that if any government official requested or subpoenaed an EFP document, word would travel like lightning throughout the farm community, threatening the goodwill established through the start-up of EFP—and the program would grind to a halt. In other words, senior government officials were warned, there was a whole lot more to be gained from a voluntary and confidential educational program that was supported at all levels of government, than from using an EFP to "fish" for violations. In the end, not only did Dr. Surgeoner receive a letter of assurance from the Ministry of the Environment and Energy (MOEE) that an EFP would not be used as evidence, the ministry also developed a document called Policy Guidelines on Access to Environmental Evaluations (November 1995). The policy clarified the circumstances under which ministry staff might apply for access to EFP documents, and aimed to strike an appropriate balance between the government's regulatory responsibilities and the right of individuals to assess their own environmental performance without fear of recrimination.[72] Dr. Surgeoner received a similar commitment from the Manager of Compliance at the Ministry of Natural Resources.

There has not been a single known case of deviation from this policy. In fact, on the handful of occasions that our Guelph office has received a call from an MOEE official about a blatant environmental infraction on a farm, the evidence was on the ground (or in the water) and the field inspector just wanted to be clear that investigators would in no way involve the farmer's EFP. We told them if the environmental contamination evidence is clear, do what you have to do.

---

72   Elliott, B. Minister, Ministry of Environment and Energy. Letter to Dr. Gord Surgeoner, Dept. of Environmental Biology, University of Guelph (Dec. 18, 1995). For the full text of the letter, see Appendix 3.

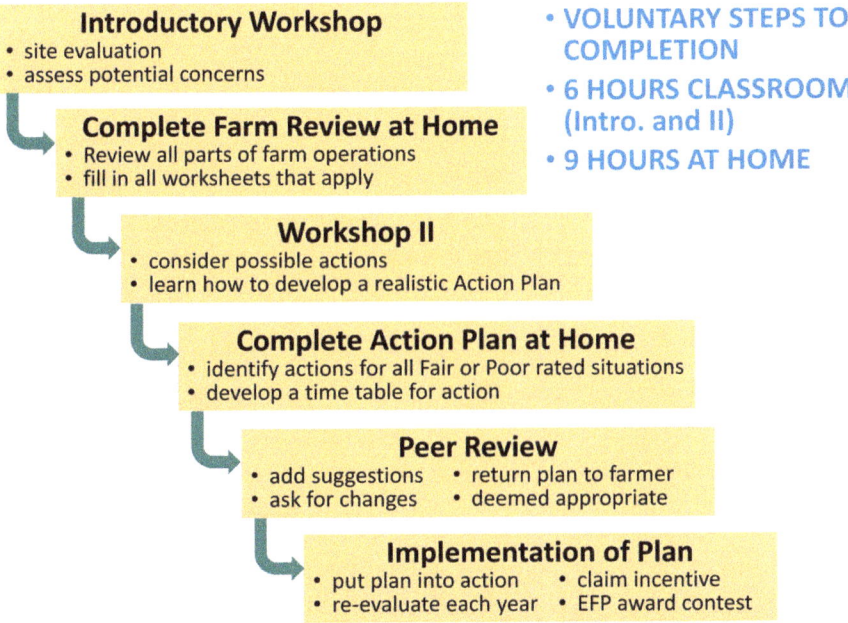

Figure 6.1 Graphic of EFP Delivery
(Source: OSCIA Archives)[73]

EFP Funding Support

The EFP program began as a pilot project in 1993 in seven selected counties/districts across Ontario. Funding for the pilot project was provided by Agriculture and Agri-Food Canada (AAFC) through the Land Management Assistance Program, with technical support provided by the Ontario Ministry of Agriculture and Food (OMAF).

In the years since, AAFC has continued to support the EFP, not only in recent federal/provincial/territorial agreements such as the Agricultural Policy Framework, Growing Forward, Growing Forward 2, and today under the Canadian Agricultural Partnership but also through the Canada-Ontario

---

73   OSCIA Archives.

Green Plan and the CanAdapt program, administered in Ontario by the Agricultural Adaptation Council (AAC). The AAC is a coalition of Ontario agricultural, agri-food and rural organizations that delivers funding and programs to stimulate growth and innovation in the farm sector.[74]

It should be noted that the EFP would not have likely survived without the loyal financial support of the AAC through seven critical years from 1997 to 2004. Funding was not guaranteed, and the delivery partners went back numerous times to justify renewed funding. OSCIA has been most grateful to the AAC staff and board of industry leaders for their continued support.

Between 2003 and 2008, funding was provided by AAFC and OMAFRA under the Agricultural Policy Framework (APF) agreement between the federal, provincial, and territorial governments. The goal of the APF was to make Canada a world leader in "food safety, innovation, and environmentally-responsible production."[75]

The role of local agribusinesses must also be acknowledged. For many EFP workshops, a local equipment dealer, seed company, financial institution, or feed elevator chipped in funds to buy lunch, coffee, or even donate their board rooms for the workshop. The culture in rural communities is one of generosity and support for each other. It also confirmed that agri-business endorsed the principles behind EFP.

OSCIA was a major beneficiary of this agreement. It was the first time that we had received a five-year commitment to support the EFP and associated cost-share funding for producers. It was also the first time since the LSP, National Soil Conservation Program and Permanent Cover programs where substantial funding in the multi-millions was provided to support a suite of programs for the agricultural communities.

---

[74] Agricultural Adaptation Council website. "A funding model that works: grassroots decision-making by the Industry for the Industry." http://www.adaptcouncil.org/aac-history.

[75] Government of Canada. "Agricultural Policy Framework: Federal-Provincial-Territorial Programs (2005)." www.publications.gc.ca/collections/Collection/A34-3-2005E.pdf (Accessed May 2, 2017).

Growing Forward

Governments are always looking for catchy names to help raise the profile of new programs and to distinguish their brand from previous governments. The years after APF were no exception. It evolved into a new generation of programming entitled Growing Forward (2008-2013) and then Growing Forward 2 (GF2) which continued to support EFP into 2018.[76]

Over the years with the growing experience and maturation of EFP, the peer review committees were discontinued in 2013 to streamline the quality control process. EFPs are now verified as complete by our highly trained workshop leaders.

Electronically Engaging

The EFP was launched just as technological advances were changing the way government programs were being administered. In the early days, our working committee supported the development of an electronic EFP so a sub-committee was formed to investigate the options. Keep in mind that in the early 1990s, electronic everything was still primitive, perhaps even fossilized by today's standards. Disk Operating Systems (DOS) were the norm, but while they were technical marvels in their day, they tended to be clunky and difficult to operate proficiently.

Our subcommittee laid out the electronic EFP requirements and submitted the task for tender. Two proposals were received, one to be built on the DOS platform and the other to be built on a new and emerging platform called Windows. The sub-committee mulled over the two options, and in the end, they concluded that since DOS was the most commonly used operating system at the time, and Windows was so new and not yet proven, it would be risky to build a Windows-based EFP. We decided to go with the DOS consultant. Wrong! The consultant was great and provided us with a satisfactory product without any scope creep. However, technology was moving so

---

76   OMAFRA website. "Canada-Ontario Environmental Farm Plan." www.omafra.gov.on.ca/english/environment/efp/efp.htm (Accessed April 5, 2017).

fast that by the time we had the e-EFP built, Windows was taking the world by storm and the DOS version was obsolete before it hit the country roads. Thankfully, the budget for this project was only $10,000, so the sting was not too painful. I'm sure we all can name data management failures that cost taxpayers much, much more. I squirm to this day when discussing data management and IT projects that are typically fraught with scope creep, expensive consultants, and complicated programming that requires continuous upgrades.

EFP Beyond Ontario's Borders

Atlantic Canada was the first region outside Ontario to adopt a similar EFP program in 1996. I was invited along with Brent Kennedy and Don Lobb to the Maritimes as part of a travelling roadshow to present the Ontario model to farm leaders, government officials, and members of the Eastern Canada Soil and Water Conservation Centre, which organized our trip. We conducted several workshops in Fredericton and Moncton, N.B., that attracted participants from Nova Scotia and Prince Edward Island. We then proceeded to Gander, N.L., where we toured the agricultural areas and met farmers to hear their perspectives on agriculture and the environment. The Eastern Canada Soil and Water Conservation Centre developed an EFP Workbook that was suitable for all of the Atlantic Provinces.

Alberta was the next province to come on board with their version of the EFP, which was launched in 2003.[77] Alberta sent a delegation of farm leaders and government staff to Ontario to meet the EFP delivery team and farm leaders. We presented a workshop on our experiences, and then took them on a tour to speak with Ontario farmers who had participated in EFP so they could hear their perspectives first-hand.

Quebec wasn't far behind. The Ontario EFP provided a French-language workbook, and the Quebec EFP model was incorporated into "les

---

77   Alberta Environmental Farm Plan — History. www.albertaefp.com/about/what-we-do#history (Accessed April 26, 2017).

clubs-conseils en agroenvironnement."[78] British Columbia test drove a pilot EFP in 2001-2002 and had a full EFP program available for farmers by 2004. Manitoba offered their EFP to producers in 2004, while Saskatchewan started in 2005. OSCIA provided data management services for the Saskatchewan EFP delivery agency, the Provincial Council of Agricultural Development and Diversification Boards, under Growing Forward from 2008-2013.

Ontario staff and farm leaders visited all of these jurisdictions to share their experiences. In all visits outside of Ontario, we emphasized the importance of farm leaders taking ownership and collaborating with their government partners. Further, we emphasized how Ontario had set up the county/district peer review committees to provide a non-threatening system of review, verification, and quality control.

Not all provinces had the luxury of an organization like OSCIA, nor did they need one. The P.E.I. Federation of Agriculture took leadership for delivery in their province. In Nova Scotia, they elected to have technical experts visit with each producer to review their conditions, identify areas of concern, and lay out an action plan to address these concerns. The one-on-one approach certainly made sense in areas where the number of farms was low and technical staff were available. Dr. Rob Gordon, Dean of the Ontario Agricultural College from 2008-2015, spent the early part of his career as one of those field staff members in Nova Scotia.

Because of our initial connection with the organizers of the Wisconsin Farm*A*Syst program, it was not long before we began networking with other American states as well. I gave presentations about the Ontario EFP program in Kansas, Iowa, and California, to name a few. Farm leaders from the Ontario Environmental Farm Coalition contributed at other locations. Through Ontario's participation in the Great Lakes Whole Farm Planning Network, each of the Great Lakes states received an EFP Workbook and heard full details about our program directly through various workshops.

---

78   Clubs conseils en agroenvironment. www.clubsconseils.org (Accessed April 26, 2017).

Outreach around the World

Word about EFP trickled out and copies of the EFP Workbook were sent to over thirty countries.[79] Of course, not all the workbook content would have been relevant within their jurisdictions. Nevertheless, the program continued to grow, and on many occasions, OSCIA was (and still is) called upon to make presentations about Ontario's EFP to the many international agricultural delegations that visit the Guelph region.

> I recall a Chinese delegation that was headed to Brazil, but planned their trip with a stopover in Toronto specifically for travel to Guelph to hear about EFP

Over the years, the U.S. government sent many delegations to Guelph and several of our farm leaders went to Washington, D.C. Bill Richards, chief of the U.S. Soil Conservation Service from 1990 to 1993 and a no-till pioneer farmer, spent the better part of a day with us, debating various soil management and conservation policies. Other countries that sent multiple delegations with a keen interest in EFP included Russia, Ukraine, and China. I recall a Chinese delegation that was headed to Brazil, but planned their trip with a stopover in Toronto specifically for travel to Guelph to hear about EFP. They told me that the EFP was the only presentation in Canada on their way to Brazil.

Among the many international delegations that visited Guelph, there was an unexpected delegation from North Korea in the early-2000s. Travelling with a church-sponsored group from Oregon, the North Koreans were intent on learning how to expand their food production. We heard tales of badly degraded soil conditions that were threatening their food supply. From the glimpses that I learned through their interpreter, it appeared that they required dramatic changes to soil management. A significant portion of their terrain is mountainous and rugged. Hay and pasture systems would have been a first step for soil restoration

---

79 OSCIA Archives. Recipients included Australia, Japan, Romania, Chile, Philippines, United States, Cuba, Finland, South Korea, China, Estonia, Spain, New Zealand, Russia, Slovakia, South Africa, Iceland, Germany, England, Ireland, Scotland, North Korea, Argentina, Uruguay, Brazil, Mexico, Costa Rica, Switzerland, Ukraine, Zimbabwe, and Yakutia.

A World-Class Environmental Farm Plan

to stabilize and improve the soil structure. Dairy and meat products, whether from cows, goats, or sheep, could provide better nutrition and protein to meet their population needs. To this day, I wonder whether their short visit with us had any impact on their soil management practices and ability to increase their food production.

Another standout was a visit by a Nuffield Scholar[80] from Zimbabwe in the late 1990s. He had sought out the OSCIA to hear about the EFP and relay some of the challenges faced by farmers in Zimbabwe, where the legacy of colonialism and government land reform policies were fuelling racial tensions. I was surprised to learn that security was the largest expense of their farm operation. Hearing about guards equipped with military vehicles and assault weapons to protect their crops and livestock on their farm was a real shocker for me. This multi-generation farmer was not optimistic about remaining on the family farmstead. The Nuffield program brought numerous visitors from around the world to learn about issues in Canadian agriculture. Nuffield Canada also supports many Canadian agricultural professionals with travel scholarships to study policies, programs, services, and marketing approaches around the world.[81]

> Among the many international delegations that visited Guelph, there was an unexpected delegation from North Korea in early-2000s

Although the EFP program attracted lots of attention from our neighbours to the south, the appearance in Guelph of another next-door neighbour caught me by surprise. No, I'm not talking about Greenland or the French territories of Saint Pierre and Miquelon off the coast of Newfoundland. Rather, it was a delegation from across the Arctic Ocean that was most memorable. Although I was invited to give a presentation to a delegation from Russia, I soon learned that the group was from the Republic of Yakutia (Sakha), or Yakutsk, in Russia's far north. Over half the territory is north of the Arctic Circle and boasts the coldest inhabited place on Earth: Oymyakon, a remote village where winter temperatures

---

80   Nuffield Canada. Grow the Leaders of Tomorrow, http://nuffield.ca/wp/.
81   Ibid.

average -50°F.[82] The delegates were also memorable because they were smartly dressed in finely tailored suits and ties, were well-educated, and had a tremendous sense of humour. I quizzed them about farming so far north and learned that breeding and grazing cattle accounted for most of the region's agricultural output. They appeared to be fairly self-sufficient, growing vegetables in the summer and harvesting bountiful fish stocks. They also were quite proud of their horses. The region is rich in mineral resources, especially diamonds. As a memento, the group presented me with a small mural coated with diamond dust. Perhaps I should get it appraised. Among the many delegations I met, the warm and spirited culture of our neighbours from Yakutia made me feel like I wanted to visit their country (in the summer of course). I would anticipate a warm welcome even if the air was frosty.

> I had not yet learned any Spanish vernacular, but my meaning was clear

As it happened, one of my most memorable journeys was a bit farther south, to Uruguay in 1999. My visit was organized by Dr. Marta Chiappe at the Universidad de la Republica de Uruguay, Montevideo. The university received a special grant from the International Research and Development Centre (IRDC) to work on an environmental farm plan suited to Latin America. I was warmly received by my hosts who arranged tours into farm country and provided a driver and translator to help me with EFP presentations along the way. In that era, presentations relied upon slides or overhead acetates as audio visual aids. At one of my first presentations, the slide projector had a horizontal slide tray, not the rotary carousel that we were accustomed to in North America. Before the presentation, I quickly transferred my slides into the horizontal holder in the correct order. Unfortunately, I was unfamiliar with some of the finer points of operating the projector, such as that if the slide holder was accidentally advanced backwards, it disengaged from the projector. I learned the hard way how not to advance slides when I hit the remote and instead of advancing the slides, the whole works ejected and spilled my slides randomly on the floor. I had not yet learned any Spanish

---

82   Weather Atlas, https://www.weather-atlas.com/en/russia/oymyakon-climate, (Accessed May 6, 2018)

vernacular, but my meaning was clear as I muttered, red-faced, a few choice English adjectives under my breath until I got the slides re-inserted in the correct order. From that point on and for the rest of the tour, the slide projector performed flawlessly. I'll leave it up to others to judge my performance. However, as we toured rural Uruguay, meetings were typically held in small-town community centres where there the local youths were fond of lighting firecrackers, sometimes near our meeting location. I was prepared to hit the floor and take cover until I became accustomed to their recreational activities.

Several years later, a colleague from the University of Guelph attended a follow-up meeting with our Uruguayan friends. An EFP suitable for conditions in Latin America was formatted and produced in Spanish. It would be intriguing to learn more about the outcomes and influences across Latin America from that investment.

Great Lakes Basin Whole Farm Planning Network

Within the Great Lakes region, environmental planning was being investigated from many different perspectives. An organization based in Minneapolis-St. Paul called the Minnesota Project took notice. The Minnesota Project had evolved initially to support alternative energy projects and networking with sustainable agriculture projects.

Led by John L. Lamb, the Minnesota Project acquired funds from the Great Lakes Protection Fund in support of environmental projects to improve water quality in the Great Lakes. The group then formed the Great Lakes Basin Whole Farm Planning Network (GLWFPN), and sought out partners in Ontario and other U.S. states in the Great Lakes watershed. The name evolved into the Great Lakes Basin Comprehensive Farm Planning Network. OSCIA participated in this network from 1995 to 1999.

This marked the first time that OSCIA knowingly received funds from a U.S. source to support our expenses related to travel and communication. It was a diverse group, with a steering committee made up of twenty-six interested and knowledgeable specialists. Ontario was represented by EFP co-ordinator Don Hill and me. One of my fond memories from that time is of a meeting held at the Max McGraw Wildlife Foundation near Dundee,

Illinois, a suburb of Chicago. McGraw was a pioneering entrepreneur and conservationist famous for, among other things, merging his electric company with Thomas Edison to form the McGraw-Edison Company. McGraw is also credited with manufacturing and marketing one of the first modern household appliances, the Toastmaster toaster.

At the McGraw facility, historically set up as a hunt camp, we were assigned narrow bunks. Various wildlife options were on the dinner menu but orders had to be placed during breakfast. A number of us had quail. I'm not sure who was assigned the task of tracking down the wild game, but we were assured it was very fresh! A quote often attributed to Max McGraw was, "There is a way to do it better… find it."[83]

The members of the Great Lakes network were following McGraw's advice. Options were being explored for comprehensive or whole farm planning that would serve to co-ordinate regulations; improve conservation and water quality; integrate economics and environment; promote sustainable agriculture; and consider quality of life.[84]

The ultimate goal was to improve farm productivity, reduce water pollution, reduce erosion, better manage nutrients, and adopt better pest management. The essential qualities of the Whole Farm Planning Process echoed the approach taken in developing the EFP in Ontario:

- The farmer is in charge.
- The farm family sets goals for the farm.
- Planning is voluntary.
- The entire farm is included.
- Problem areas are clearly identified.
- Alternate options are considered.

---

83 "Max McGraw's Life Story." Max McGraw Wildlife Foundation Website. http://www.mcgrawwildlife.org/mcgraw-history (Accessed April 26, 2017).

84 Kemp, L. "Successful Whole Farm Planning: Essential Elements Proposed by the Great Lakes Basin Farm Planning Network." The Minnesota Project (July 1996, OSCIA Archives).

- The farmer develops an action plan with appropriate timelines.

- Implementation progress is measured and the plan is intended to be revisited.

- The planning process is encouraging, easy to understand, educational, and technical assistance is available.

- The farm plan itself is confidential.[85]

> Nancy Grudens-Schuck, a PhD candidate from Cornell University who went on to a faculty position at Iowa State, earned her PhD by conducting participatory action research (PAR) on EFP. She moved her family to Guelph, Ont., for a full year from 1995 to 1996 for part of her research

The Minnesota Project team supported the Great Lakes Basin Comprehensive Farm Planning Network to deepen and strengthen outreach among farm groups, increase the number of whole farm plans completed in the basin, and help evaluate and educate farmers and agency personnel about the different types of whole farm planning models. As part of those efforts, it published stories about whole farm plans in six of the eight states and Ontario.[86]

EFP: The Most Researched Program Ever!

Participation in the Great Lakes network led to several other significant collaborations. Nancy Grudens-Schuck, a PhD candidate from Cornell University who went on to a faculty position at Iowa State, earned her PhD by conducting participatory action research (PAR) on EFP. She moved her family to Guelph, Ont., for a full year from 1995 to 1996 for part of her research.

I travelled with several colleagues to Cornell for her dissertation in August 1998. Using ethnography research methods, her research included engagement

---

85 Ibid.
86 "Great Lakes Basin Comprehensive Farm Planning Network: Phase II Implementation Timeline." Great Lakes Protection Fund, www.glpf.org/funded-projects/great-lakes-basin-comprehensive-farm-planning-network-phase-ii-implementation/ (Accessed April 26, 2017)

with our provincial committees, county EFP workshops, staff, and other management activities.[87] She wrote that the EFP experience in Ontario was a useful model for studying how to stimulate and manage meaningful debate when stakeholders may disagree about the task at hand and who should provide leadership to address the issues.

Researchers closer to home were also keen to study the EFP. At one point in time, I had a whole filing cabinet drawer full of consultant's reports, master's theses, and evaluations with which OSCIA had been involved. For my own master's thesis,[88] I listed and summarized thirteen such research studies in an appendix. Research is continuing, specifically related to how an EFP may fulfill the environmental component of a sustainable farm and food plan. The links between EFP and sustainable food are discussed in more detail in Chapter 13.

Drinking Water Protection in New York State

Through our connections with the Great Lakes network, we kept hearing good news about how farmers were working with numerous agencies to protect drinking water for large urban centres such as New York City (NYC) and Syracuse. We also heard about generous subsidies that New York State farmers received to invest in water protection measures, such as fencing to keep livestock away from watercourses and manure storage structures to hold manure for application in the fields when soil and weather conditions were ideal. Conservation planting equipment was also funded by state water quality improvement programs to eliminate soil erosion.

We wanted to learn more so in September 1996, we organized a bus trip to see first-hand the details about the New York State programs. We filled the coach with fifty people including farm leaders, extension staff from the

---

87 Grudens-Schuck, N. "When Farmers Design Curriculum: Participatory Education for Sustainable Agriculture in Ontario, Canada." (August 1998) https://www.researchgate.net/publication/34052949_When_farmers_design_curriculum_participatory_education_for_sustainable_agriculture_in_Ontario_Canada (Accessed April 27, 2017).

88 Rudy, H. "Performance Measures for Environmental Programs Using the Environmental Farm Plan as the Basis for Analysis." (OSCIA Archives).

agriculture ministry, conservation authorities, the Ontario Federation of Agriculture (OFA), and OSCIA. We visited a large dairy family farm that milked several thousand cows. We also spent time at a small dairy farm high up in the Catskill Mountains, not far from the large reservoir that supplied a significant portion of New York City's drinking water.

In his report back to colleagues at the OFA, David Armitage, the organization's director of regulatory reform, wrote that New York's approach to managing agriculture's impacts on watersheds offered some valuable lessons for Ontario. By protecting urban water supplies at the source and investing in comprehensive farm planning in the watershed, the state managed to avoid spending billions of dollars on a filtration plant for New York City that otherwise would have been required by federal regulations. Instead, the Watershed Agricultural Program costs NYC ratepayers about US$20,000 a day, compared to the US$1 million per day it would cost to operate a filtration plant on top of US$6-8 billion needed to build a plant with the capacity to meet New York's needs.[89]

> By protecting urban water supplies at the source and investing in comprehensive farm planning in the watershed, the state managed to avoid spending billions of dollars on a filtration plant for New York City

Although successful in a strict cost-benefit sense, the New York model had several drawbacks. Farmers weren't involved in developing and delivering the program and there was little incentive to participate beyond accepting the advice and cashing the cheques being offered up by experts from the big city. For farms that were eligible, the program provided up to US$100,000 for capital improvements with zero investment required by the farmer. As a result, the program provided a lot of money to relatively few farms—as well as a small army of technical experts and construction contractors—for projects that sometimes made little economic sense. According to David

---

89   Armitage, D. "Report to Ontario Federation of Agriculture Executive and Environmental Committees." (Oct. 16, 1996) Personal communication between D. Armitage and H. Rudy, OSCIA (Sept. 8, 2017). The full text of Armitage's report is reprinted in Appendix 4.

Armitage, "the Ontario EFP model of using limited resources to influence a far greater number of farmers is more broadly applicable."[90]

Another intriguing take away from that tour was the small line-item for drinking water protection on New York City residents' water bills. It was a modest amount collected from each citizen, but a huge total sum to invest in helping farmers to upgrade their facilities and improve land management near drinking water reservoirs. As a result, each day, more than 1.1 billion gallons of fresh, clean water is delivered from the Catskill Mountains and upstate reservoirs to the taps of nine million customers. Over 97 per cent reaches homes and businesses through gravity alone.[91]

Similar farm support programs were in place throughout the Lake Skaneateles watershed, which provides drinking water for the city of Syracuse. The Skaneateles Lake Watershed Agricultural Program was established in 1994 to support environmentally sound farming practices where they targeted pathogens, nutrient management, and soil erosion. The philosophy was simple: protecting water quality at the source by providing farmers with financial support to adopt protective measures was much more cost-effective than installing a costly filtration system. [92] Appendix 4 contains the full report submitted by David Armitage.

Interestingly, I returned to the town of Skaneateles a few years ago on vacation. When walking out onto the pier at the lake on the edge of town, I saw numerous plaques attached to pier posts describing the good work of the Skaneateles Lake partners in protecting the lake's water quality. In 1996, our group from Ontario was impressed with the commitment of citizens and farmers to protect their drinking water. These watershed management programs are applicable today as a model to guide and protect drinking water in many other regions around the world.

---

90   Ibid.

91   NYC's Reservoir System. www.nyc.gov/html/nycwater/html/drinking/reservoir.shtml (Accessed April 27, 2017).

92   Cooperative Conservation America. http://cooperativeconservation.org/viewproject.aspx?id=87 (Accessed April 27, 2017).

Last Generation Farms

I had not heard the term "last generation farm" before our 1996 trip to New York State. One dairy farmer described how some family-owned farms were disappearing once the current owners retired or moved away—they were being rented by larger farms, or sold to people attracted to the rural lifestyle but who had no interest or expertise in farming. Although it does not show up commonly in the literature, the term perfectly describes the trends in agriculture in recent decades. It is difficult for a small farm to survive economically and provide a standard of living on par with that enjoyed by urban residents. Family farms in North America have been forced to expand in size to achieve economies of scale. The largest farms are still family owned, but there may be several generations of families working together in order to achieve an adequate standard of living. In Canada, 97 per cent of farms are family owned. [93]

Farm Credit Canada suggests there are two major reasons for the decline in farm numbers:

- Increasing competitive pressures for more acres. Value within an operation can be generated through production efficiencies and the resulting lower costs, which are partly a function of size. This is certainly true for grain and oilseed operations producing a standardized commodity.

- Technological advancements. Larger, more efficient equipment covers more acreage. Typically, the adoption of new technology

---

[93] 2016 Census of Agriculture. Statistics Canada (May 10, 2017). www.statcan.gc.ca/daily-quotidien/170510/dq170510a-eng.htm (Accessed Aug. 8, 2017).

decreases the per-acre costs of production—especially (but not strictly) when the number of acres also increases.[94]

Meanwhile, consumers in industrialized nations have benefited from modern agriculture's economies of scale, enjoying convenient access to high-quality food at historically low cost. In Canada in 2015, average spending on food including restaurants accounted for just 10.4 per cent of household income. [95]

Does our society care about the shifting demographics in our rural communities? Certainly, larger farms generally have the cash flow and capital reserves to invest in environmental improvements. Although it is not OSCIA's current mandate to engage in debate on land use policy, two decades later I still think on that conversation about last generation farms. Has it been wrong for our farmers to become more efficient over time by shifting to larger and larger operations? The consumer does win! What about new farmers wishing to start out farming? How do they compete, or even afford the land and equipment needed to get started? Is there something about our food policy or our land use regulations that require fixing?

EFP Goes Down Under

The presence of armed guards at the airport terminal alarmed me as I pinched myself awake after the ten-hour flight from Toronto. As we made our way through the terminal toward the baggage carousel, I was not expecting to see machine-gun-toting police. After all, this was still the United States, not some police state. I was in Honolulu, Hawaii, on a stopover en route to Sydney, Australia, just over a month after the Sept. 11, 2001, terrorist attacks on the World Trade Center. Post-9/11 air travel was still unsettling. To see beefed up security on the ground reminded me that we had entered a new era for air travel.

---

[94] "What's Influencing Canadian Farm Numbers in 2016?" Farm Credit Canada. (May 9, 2017). www.fcc-fac.ca/en/ag-knowledge/ag-economics/whats-influencing-canadian-farm-numbers-in-2016.html (Accessed May 23, 2017).

[95] "Survey of Household Spending, 2015." Statistics Canada. www.statcan.gc.ca/daily-quotidien/170127/dq170127a-eng.htm (Accessed May 23, 2017).

A World-Class Environmental Farm Plan

The Australian trip had been months in planning after Bruce Lloyd, a dairy farmer and chair of the Australian Landcare program, visited Ontario where we introduced him to the EFP program. I had guided him on a personal tour of the countryside to meet farmers and staff engaged in EFP. Through our conversation, I learned that an Australian environmental program, Landcare, had become a national force just a few years after it was launched by the Victorian Farmers Federation in 1986. Today, the Landcare movement is made up of more than 5,400 groups across Australia.[96]

> Australia and Canada have much in common: each has a very large land mass and low population, with an export-driven economy heavily dependent upon agriculture and natural resources

As non-profit initiatives, EFP and Landcare had similar features. They were both locally driven, focused on issues that mattered most in their communities, and each sought support from funding partners and the public. Unlike EFP, Landcare was more inclusive in that it brought together community groups outside the farm community for conversations about local priorities. In my time with Mr. Lloyd, we discussed at length how our programs could evolve into an environmental management system (EMS) that increased the level of rigour to improve accountability and help farmers establish a market advantage by demonstrating their commitment to sustainable practices. Australia and Canada have much in common: each has a large land mass and low population, with an export-driven economy heavily dependent upon agriculture and natural resources. We could learn more from each other and our conversations resulted in the invitation for me to visit Australia.

Mr. Lloyd was impressed to learn that by the summer of 2001, some 20,000 Ontario farmers—representing close to half the province's tillable acres—had voluntarily enrolled in the EFP program. A financial incentive of $1,500 per farm business provided by the federal government ($11 million in total) had leveraged close to $60 million in on-farm investments to address

---

96   Landcare Australia. www.landcareaustralia.org.au/about/ (Accessed May 1, 2017).

environmental concerns. Many other provinces in Canada had adopted or were in the process of adopting a similar program. The wheels were in motion for me to visit Australia in October-November.

My itinerary took me to workshops and on-farm visits throughout the states of Victoria, New South Wales and Queensland. I provided an overview of the Ontario EFP and learned about their local environmental challenges. Preventing salinization of their soil, protecting wildlife, and stewardship of the natural environment were high priorities for Australian farmers. There was ample time for discussions around how producers might gain a marketing advantage by improving their environmental performance. Key organizers included Philippa Rowland, a senior scientist with Australia's department of agriculture, forestry, and fisheries. Tony Pexton, deputy president of the National Farmers Union in the United Kingdom, joined our tour and conferences. Mr. Pexton was farming 1,000 acres devoted to wheat, barley, oilseed rape, vining peas, and hogs. He provided insight on British EMS programs.

Genevieve Carruthers of the Wollongbar Agricultural Institute organized a key conference in Ballina, New South Wales. Hospitality was second to none. I was met at the airport by Geoff McFarlane, project leader for Environmental Best Management Practices (EBMP) with Victoria's department of natural resources and environment, who toured me throughout the state. A highlight took us along the Great Ocean Road, stopping at farms along the way as well as typical tourist attractions. Geoff returned to Ontario in 2003 to once again compare notes. Thanks to Geoff's persistence, the EBMP had developed self-assessment characteristics similar to EFP's but adapted to Australian conditions.

The Australia visit was certainly one of the highlights of my OSCIA career. To arrive in another country and not worry about any transportation, food or accommodation for three weeks was indeed a pleasure. Interestingly, upon my departure, Ms. Rowland not only took me to the airport, but walked with me right into the departure lounge, no questions asked, as I boarded for Canada. No security there. The impact of 9/11 had not yet reached that small airport Down Under.

I gratefully acknowledge the generous support of the many sponsors who made the 2001 Australian EMS study tour possible.[97]

Upon my return to Canada, I provided my sponsors with a report. Many discussion points share similarities with current Canadian efforts on sustainable farming. I include the Executive Summary and Recommendations in Appendix 5 because I believe these points are just as relevant today, not just for Canada or Australia but for any other jurisdiction engaging in discussions about environmental management systems.[98]

OSCIA certainly can't take sole credit for the EFP program's success. The Ontario Farm Environmental Coalition, led by the Ontario Federation of Agriculture, was key in the political process and assisted in acquiring funding. The technical content, which provided credibility for best management practices, was the responsibility of OMAFRA staff who also tag-teamed with OSCIA at EFP workshops. Kudos to AAFC staff for the support from various funding programs to keep EFP available to producers for over twenty-five years. OSCIA has been very fortunate to be the delivery and administrative arm for EFP across Ontario.

Expanding OSCIA's Outreach to Communities in Ontario

> *"OSCIA is a fantastic organization with real connections to the grassroots farmers across the province of Ontario. OSCIA has an excellent reputation for reliable and dependable delivery of various programs to those in rural communities. It was an absolute privilege to serve as OSCIA president."*
>
> *-Steven Eastep, 2004 OSCIA President (Wellington County)*

---

97   My tour was sponsored by: the Grains Research and Development Corporation (GRDC); Australian Landcare Council; Ballina EMS Conference; Victoria Department of Natural Resources and Environment (DNRE); Wollongbar Agricultural Institute; Ricegrowers' Association of Australia; Environmental Best Management Practices on Farms Project, and the Department of Agriculture, Fisheries & Forestry – Australia (AFFA).

98   Rudy, H. "Australian Environmental Management Systems: A Canadian's Perspective — Study Tour through Southeastern Australia, October 24-November 14, 2001." (OSCIA Archives)

The EFP opened doors in communities where OSCIA previously had little engagement; for example, among the many distinct groups of Old Order Mennonites and Amish who call Ontario home. From their Ontario roots in Waterloo County, these communities can now be found in far-reaching clusters from the Ottawa Valley to the Algoma, Sudbury, Temiskaming, Cochrane, and Rainy River Districts, where they've been drawn by lower priced, readily available farmland. Their reasons are similar to the factors that motivated their ancestors who followed the trail of the black walnut trees from crowded Pennsylvania to the rich farmland of Waterloo County at the beginning of the nineteenth century.[99]

> Richard knew that if the local bishop could be convinced of the benefits in attending an EFP workshop, the followers would participate

Often referred to as the "horse and buggy people" or "plain people," these communities have historically shied away from public meetings and government-sponsored workshops. Nor did many obtain government grants for improvements on their farms. However, EFP opened doors in many of these communities by raising awareness about common environmental concerns. Their beliefs would naturally support being good stewards of the land and what better way to learn how to be good stewards than to attend an EFP workshop to learn about best practices? OSCIA's EFP workshop leader in Waterloo Region had deep roots in the Mennonite culture. Richard Lichty understood these dynamics and was instrumental in meeting with various bishops to discuss options suitable to their congregations. Richard knew that if the local bishop could be convinced of the benefits in attending an EFP workshop, the followers would participate. Before Richard's untimely death in 1998, he led EFP workshops in the Wellesley and East Perth municipalities where black buggies lined the parking lots. Frank Kains, a former OMAFRA engineer, continued Richard's work as the region's next EFP workshop leader.

---

99   Bloomfield, E. "Building Community on the Frontier: The Mennonite Contribution to Shaping the Waterloo Settlement to 1861." The Newsletter of the Mennonite Historical Society of Ontario. www.mhso.org/sites/default/files/publications/Ontmennohistory15-2.pdf (Accessed Sept. 18, 2017).

Some communities would participate in the EFP workshops to learn about best land stewardship practices, but not take part in the cost-share components. Other communities cleared their conscience to accept grants for on-farm environmental improvements, typically for fencing to keep livestock from waterways, or costly manure storage infrastructure to eliminate nutrient runoff. Additionally, through the efforts of outreach partners such as Anne Loeffler at the Grand River Conservation Authority, some communities received their conservation authority financial support to help keep rivers and streams clean.

> EFP also opened up communications with First Nations communities

As Mennonite communities expanded across Ontario, similar rapport was established by EFP workshop leaders such as John Benham (Wellington County), Ray Robertson (Grey County), Lois Sinclair (Huron County), and Mary McIntosh (Perth County), to name a few. These communities are engaged in adopting best land stewardship practices, regardless of their acceptance of government grants. Through eastern Ontario and across the north, Mennonite farmers are investing in improvements to their expanding land base by incorporating best stewardship practices with tile drainage, crop rotation, and building repair and construction.

EFP also opened up communications with First Nations communities. With the rapport established under EFP, clients from the Indian Agricultural Program of Ontario (IAPO) expanded their participation. Workshop leaders Pam Charlton, Joanne Sanderson, Mary Scott, and Margaret May also facilitated Growing Your Farm Profits (GYFP) workshops designed specifically for IAPO clients across the province.

Margaret May, the former EFP workshop leader for Middlesex and Elgin counties who is now program lead in southwestern Ontario, reflects on those experiences:

"I had some First Nations producers attend EFP workshops in Middlesex early in the process. I did facilitate a dedicated EFP workshop for the Moraviantown Delaware Nation at their request. The lead grower on Walpole Island also attended a workshop that I facilitated, as the island is farmed as one entity. We discovered these partnerships were far more valuable once a connection had been established. My mother taught school with the Oneida Nation of the Thames and Munsee Delaware Nation and this link was incredibly beneficial. The link with my mother as teacher was verified by the band chiefs. Later, when we were looking for on-farm research participants with the OSCIA-Dekalb Roundup Ready Corn Showcase, some of the co-operators were farmers from these communities. The link established through that research helped them accept the EFP process as well. Like most programs, word of mouth is absolutely the best promotion we can get.

"EFP is a relatively easy sell to the producers: assess your business, plan for changes and apply for funding to help to make the upgrades. The concept of GYFP was a bit more abstract. IAPO was good to encourage producers to attend to upgrade their record keeping processes and they (IAPO) provided one-on-one help to advance the record-keeping skills of their farmers.

"I led Walpole Island and Six Nations through a modified version of Growing Your Farm Profits to assess the business skills of the band councils in both of those areas—these worked well. The notion of succession planning caught their interest. It wasn't in the traditional sense, but they were keen to start succession planning for their farm manager position. Their identified successor was accepted to the Advanced Agricultural Leadership Program and is now acting as assistant Farm Manager—a good news story for sure!

"I still have regular contact with some of the Moraviantown growers and the farm at Walpole Island as well as some from Six Nations. They are keen to keep up with programs and new developments!"[100]

EFP also attracted participants from the six First Nations communities on Manitoulin Island: Wikwemikong, Sheguiandah, M'Chigeeng, Aundeck Omni Kaning, Zhiibaahaasing, and Sheshegwaning. According to Mary Scott, workshop leader on Manitoulin, OSCIA has worked with First Nations since EFP workshops began in that area in 1994.

"Participants have come to attend workshops from the various First Nations communities and applied for cost-share programs over the years. We

---

100   Personal communication between Margaret May and Harold Rudy (June 26, 2017).

have regularly held workshops for EFP and GYFP in Wikwemikong First Nation.

"I have attended the agricultural information meetings held in Wikwemikong First Nation in order to inform their producers about the OSCIA programs available. In the past, we have worked with Mark Leahy of the Indian Agriculture Program of Ontario (IAPO) to promote and deliver programs.

"Manitoulin OSCIA annual meetings are attended by First Nations members with some of their members serving on the Manitoulin Soil and Crop Improvement Association board of directors."[101]

> Advances in plant genetics will open up new frontiers for agriculture, making it possible to grow new varieties of drought-resistant, frost-tolerant, and disease-resistant crops in improbable places, not just in Canada's north but around the world

There you have it. Field staff are directly connected to engage local communities. In one of my recent meetings in Sudbury, I had been informally approached about potential soil and crop improvement work in communities as far north as Moosonee. With global warming and technological advancements in the offing, it would seem logical that food growing opportunities may one day abound in Canada's far north. Imagine the cold tolerance and frost-resistant traits that could be developed for food crops using some of the new gene-editing techniques.[102]

With proper scientific scrutiny and regulatory controls, new technologies may sprout solutions to some of the world's most urgent problems. Advances in plant genetics will open up new frontiers for agriculture, making it possible

---

101 Personal communication between Mary Scott and Harold Rudy (June 28, 2017).
102 "Precise Gene Editing in Plants." MIT Technology Review. 10 Breakthrough Technologies. https://www.technologyreview.com/s/600765/10-breakthrough-technologies-2016-precise-gene-editing-in-plants/ (Accessed July 21, 2017).

to grow new varieties of drought-resistant, frost-tolerant, and disease-resistant crops in improbable places, not just in Canada's north but around the world.

Best Management Practices: A Series

Running parallel to EFP was the development and delivery of Best Management Practices (BMP) publications, an award-winning series published in both official languages that illustrates the expectations for managing farms while conserving or improving the surrounding environment. Many BMPs profile soil management systems and how-to techniques. Although EFP defined best practices expectations, the BMP publications provide an in-depth look at each issue with colourful photos and graphics.

The goal of the BMP series has been to:

- Present affordable options for protecting soil and water resources on the farm;
- Support individual farm planning and decision-making in the short and long term;
- Harmonize productivity, business objectives, and the environment;
- Present a range of circumstances and options to address a particular environmental concern. Farmers were encouraged to use the information to assess what's appropriate for their own property.[103]

Although OSCIA did not head up the BMP development, staff members including Andrew Graham and key farmers were engaged as resource experts, working along with researchers, extension staff, and agribusiness professionals. The Ontario Federation of Agriculture (OFA) was a key administrator, channeling the financial resources to support the team that was headed by Ted Taylor, OMAFRA's BMP Publications Program Adviser. Kudos to the original BMP team formed in 1989, which also included the expertise of Adam Hayes, Anne Verhallen, Christine Brown, Lisa Cruickshank, Keith

---

103   Best Management Practices Series, OMAFRA, http://www.omafra.gov.on.ca/english/environment/bmp/series.htm#3 (Accessed January 18, 2018).

Reid, Bob Stone, Brent Kennedy, Jim Myslik, Don Hilborn, Cecil Bradley, and Gary Nelson.

EFP workbooks and the accompanying BMP publications are an in-depth resource for awareness to assist producers to address a broad array of environmental concerns. However, an exclusive opening for OSCIA arose in 1991 with special emphasis on safe drinking water for farm families, their livestock, and drinking water protection for their neighbours. This unique opportunity for drinking water investigation is outlined in the next chapter.

Chapter 7
LOOK OUT BELOW YOUR FEET

In the early 1990s, many concerns were expressed about the impact of agricultural practices on groundwater in general, and drinking water on farms specifically. I served on the advisory committee of two hallmark groundwater surveys that were part of one of the most comprehensive well water studies in North America. Funding for this comprehensive analysis flowed through OSCIA's office. OSCIA field staff were engaged in obtaining many well water samples from farms. Thanks to Agriculture and Agri-Food Canada (AAFC), these projects received sufficient funding to conduct a thorough evaluation.

> Thanks again to the strong leadership of the co-chairs, lemonade was made from the lemons

Led by Dr. David Rudolph of the Waterloo Centre for Ground Water Research at the University of Waterloo, and Dr. Michael Goss, University of Guelph researcher with the Centre for Land and Water Stewardship, a working group brought together representatives from OSCIA, AAFC, universities, and the Ontario ministries of agriculture, environment, and health. The group was anxious to get a water sampling project underway with $500,000 made available from OMAFRA and AAFC through the federal-provincial Environmental Sustainability Initiative. The only wrinkle was the project had to be completed in less than a year and the funds spent by March 31, 1992.[104]

Ideally, a well water survey should have been done in the summer when there were lots of field activities underway on the farm (potential to find bacteria, nutrients, or pesticides), but the summer of 1991 crept by and we were suddenly into September with nothing yet organized. Thanks again to the strong leadership of the co-chairs, lemonade was made from the lemons. We would conduct a winter survey, and then, if we could find the funds the following year, we would conduct a sampling at the same locations during the summer.

Using a sophisticated layering technique targeting dominant soil types, land use activities and intensity of farming, maps were drawn up to pinpoint farms on the targeted soil types. OSCIA field staff were asked to visit the communities identified on the map and find the closet active farmer willing to participate. Thanks to their success in delivering previous programs, field staff enjoyed the credibility and trust of farmers who were assured that any information they provided would be kept confidential.

In summary, there were two components of the project:

1. Well water samples were taken from 1,300 domestic farm wells.

2. Multi-level well (MLW) monitoring by technicians and scientists was carried out to enable sampling at discrete depth levels, providing data needed to create a vertical water quality profile. Samples were sent to a laboratory to check for nitrate-N[105], total and fecal coliform bacteria, and several common herbicides.

---

104 Canada-Ontario Environmental Sustainability Initiative (ESI) 1991-1992. "The Agreement." www.agrienvarchive.ca/esi/esiagree.html (Accessed Jan. 25, 2017).

105 Nitrate-N or nitrate nitrogen refers to the concentration of nitrogen present in the water that may be due to seepage from the use of certain fertilizers in the soil, or from decomposing organic matter and human waste. Nitrates are beneficial to plants, but too much nitrate nitrogen in drinking water can cause illness and even death in human infants.

Note that drinking water objectives, standards, and guidelines are terms that are sometimes used interchangeably. An explanation of Ontario's Drinking Water Quality Objectives can be found online.[106]

Results of the project showed several concerning trends:

- "37 per cent of all wells tested contained one or more of the target contaminants at concentrations above the provincial drinking water objectives.
- 31 per cent exceeded the maximum acceptable concentration (MAC) for coliform bacteria.
- 20 per cent had fecal coliform bacteria.
- 13 per cent exceeded the MAC for nitrate (seven per cent exceeded the maximum acceptable concentration for both coliform bacteria and nitrate; six per cent exceeded the acceptable concentration of nitrate alone).
- Eight per cent had detectable levels of pesticides, and one well showed pesticide concentrations exceeding the interim maximum acceptable amount.

Of 160 water wells tested (for petroleum), no petroleum-based derivatives were detected."[107]

Other key findings:

- Problems were more likely with wells that were more than sixty years old.
- Shallow bored or dug wells showed more frequent contamination.

---

106   Ontario Ministry of Environment and Energy, Water Management: Policies, Guidelines, Provincial Water Quality Objectives, https://www.ontario.ca/page/water-management-policies-guidelines-provincial-water-quality-objectives#section-13 (Accessed January 9, 2018).

107   Ontario Farm Groundwater Quality Survey Winter 1991-1992. (Sept. 1992). Executive Summary. www.agrienvarchive.ca/download/ofgqs_winter_91-92.pdf (Accessed Feb. 15, 2017).

- Nitrate-N concentrations were higher in more permeable soils such as sand.

- Coliform bacteria increased as the distance to feedlots decreased.

- Nitrate—N levels were more serious in areas used to grow crops such as corn that require a lot of nitrogen-based fertilizer.

The MLW monitoring involved installation of 144 test wells in fields and woodlots in intensively farmed areas, mostly in sandy or permeable soils. This portion of the study was to determine how far nutrients, bacteria, or pesticides moved into the soil. An interesting finding was that the frequency of occurrence of coliform bacteria concentrations above acceptable limits were similar in both the field and woodlot settings.[108]

However, the job was not done with just one round of samples during the winter of 1991-1992. In true scientific fashion, the task had to be replicated. Thanks to the federal-provincial Land Management Assistance Program (LMAP), $750,000 was made available to contract with OSCIA for another round of tests over the summer of 1992.[109]

The follow-up study involved re-sampling about 1,300 domestic farm wells and analyzing concentrations of nitrate-N, total and fecal coliform bacteria, and several common herbicides. The majority of the wells (900) were located in areas of intense agriculture with the most common soil types and on farms involved with the most common agricultural land-use practices. The remainder of the wells were located in less agriculturally intense areas including northern Ontario.

In October 1992, the Ontario Ministry of Health revised the limits for maximum allowable concentrations (MAC) of total coliform bacteria in private drinking water supplies. The new MAC was five colonies per 100 ml, down from ten colonies per 100 ml. The water samples were analysed relative to both the old and new standards. The follow-up study found:

---

108 Ibid.

109 Land Management Assistance Program. The Agreement, Schedule A. www.agrienvarchive.ca/lmap/lmapagre.html (Accessed Jan. 25, 2017).

- "40 per cent of all wells tested contained one or more of the target contaminants at concentrations above the previous provincial drinking water objectives (43 per cent with the updated objectives. (Note: Objectives at that time refers to measurable standards for water quality.)

- 32 per cent exceeded the previous MAC for at least one of the coliform bacteria selected for analysis (36 per cent with the updated MAC).

- 25 per cent had fecal coliform bacteria.

- 15 per cent exceeded the MAC for nitrate (seven per cent exceeded the MAC for both coliform bacteria and nitrate).

- 12 per cent had detectable levels of pesticides; two wells showed pesticide concentration in excess of Ontario interim maximum acceptable concentration (IMAC) values".[110]

Where the MAC of any contaminant was exceeded, the landowner was notified and encouraged to take appropriate measures to ensure the water was safe for their family and livestock. In fact, at most EFP workshops, there was strong emphasis on best practices to improve the safety of drinking water for farm families, their livestock, and neighbours. Many of the cost-share programs administered by OSCIA over the past three decades provided support to farmers for well upgrades and improvements. Many conservation authorities offered similar assistance.

As a benchmark groundwater study, this project provided an excellent baseline from which to compare for future studies. As a measure of performance over time, what would be the results if a similar survey and analysis was conducted today? The Working and Advisory committees are listed in Appendix 6.

---

110   Ontario Groundwater Quality Survey, Summer 1992 (June 1993). Executive Summary, p. 1. www.agrienvarchive.ca/gp/download/ofgqs_93.pdf (Accessed Feb. 15, 2017).

Chapter 8

THE CASE OF THE MISSING BARN OWL

I like owls. Our kids like owls and our grandchildren like owls. In 2016, we took our grandkids to the Butterfly Conservatory near Cambridge, Ont., to see a live owl exhibit. There were also some (deceased) stuffed owls and even a table displaying owl parts including wings, feet and beaks. Three-year-old Heidi stood bug-eyed in front of the live birds, while her twin brother Parker couldn't get enough of touching and hamming it up with the owl parts spread out on the table in front.

Barn owls are an endangered species in Ontario and populations across Eastern Canada were listed in 2016 as critically imperiled. The barn owl is found on every continent except Antarctica, but because it is a warm-climate species, it has been found only occasionally in Ontario along the shores of Lake Erie and Lake Ontario.[111]

On our farm, we have all the ingredients, except one, to encourage barn owls. We have an old bank barn that has a few holes under the eaves, a hay field next door and forty acres of natural landscape, a bridge within 500 metres and a river running through the property, all integral for good habitat to attract, feed, and protect barn owls. Their graceful gliding and noiseless flight would be a great delight to see. The missing ingredient is location. Our farm is marginally too far north for their liking. What will happen with

---

111 "Recovery Strategy for the Barn Owl (*Tyto alba*) in Eastern Canada." Species at Risk Public Registry. Government of Canada. www.sararegistry.gc.ca/document/default_e.cfm?documentID=2891 (Accessed May 5, 2017).

climate change? If I live long enough, perhaps I'll spot my first barn owl in the back forty.

OSCIA has administered numerous stewardship programs for species at risk, but not without some controversy.

Farmers express concerns about how species at risk (SAR) regulations may affect their management choices. In 2014, OSCIA undertook a study to engage farmers in this discussion. The results were as follows:

"An overwhelming majority of the Ontario farmers surveyed care about the environmental health of their land (96 per cent) while 77 per cent state that they care about protecting SAR. Moreover, 59 per cent would like to learn more about how they could better support SAR and 62 per cent feel that the current policies in place to protect SAR provide a benefit to society. On the whole, however, the survey results show that awareness of specific currently listed SAR is quite low among the respondents."[112]

"Results of the focus group and a larger survey [of] the responding farmers show that SAR may pose restrictions to farming operations or land uses under current legislation. Further, farmers are unclear what it might mean if at-risk species are found on their land or how current SAR legislation may affect them. Seventy-eight per cent of respondents agree that farmers have more of a burden to bear when it comes to protecting SAR in comparison to other private landowners. Financial setbacks may occur if farm activities are restricted or valuable land taken out of production due to the presence of SAR. Although 39 per cent of respondents state that they are aware of how the Ontario Endangered Species Act may affect them, almost half (43 per cent) are unsure of how the act impacts them or their agricultural activities."[113]

OSCIA first became engaged in supporting habitat for wildlife through a project entitled Managing Agricultural Drains to Accommodate Wildlife.

---

112   Garrah, L. "Risk or Reward: An Investigation of Ontario Farmer Perceptions of Species at Risk." OSCIA. (April 2014). Executive Summary. www.ontariosoilcrop.org/wp-content/uploads/2015/08/sar_survey_results.pdf (Accessed Feb. 15, 2017).

113   Ibid.

The Case of the Missing Barn Owl

Funded by Agriculture and Agri-Food Canada (AAFC), the project ran from 1993 to 1997.

Agricultural drains are of incredible importance to the viability of Ontario agriculture. Historically, whether agricultural or urban, drains were seen as a conduit to transfer surplus water as quickly as possible with little concern for habitat within the waterways. Headed up by Andrew Graham, now OSCIA's executive director, this project brought agricultural drainage interests and wildlife interests together.[114]

> A key objective was to demonstrate how the Drainage Act could support new ideas to enhance fish habitat and wildlife along these drains without sacrificing the integrity of the drain function

The project profiled improvements to four locations across Ontario:

1. South Branch of the South Nation River Municipal Drain (South Nation Conservation Authority)
2. Halls Creek (Upper Thames River Conservation Authority)
3. James Berry Municipal Drain (Township of Norfolk)
4. Cranberry Creek Municipal Drain (Township of Norfolk)

A key objective was to demonstrate how the Drainage Act could support new ideas to enhance fish habitat and wildlife along these drains without sacrificing the integrity of the drain function. Here are some examples of enhancements that might appear in an engineer's report under the Drainage Act that would also satisfy regulations in the federal Fisheries Act:

- **Boulder and rock weirs**: installed into streams to divert water flow for resting areas and food sources for fish;
- **Wooden bank covers**: wooden timbers and planks installed under water along the banks to create cover and protection for fish;

---

114 Managing Agricultural Drains to Accommodate Wildlife. OSCIA brochure. (OSCIA Archives).

- **Aquatic vegetation establishment**: restore native vegetation and remove invasive plants to provide food and cover for fish;

- **Bypass structures**: channels which allow migrating fish to move around obstacles such as dams or water-control structures;

- **Natural channel design**: creation of pools, riffles, and low-flow channels to provide refuge for fish, habitat diversity, and enhance flow during low-flow seasons;

- **Sand traps**: excavated depressions which collect silt to reduce its movement downstream may reduce cleanout costs while minimizing disturbance during spawning;

- **Retention ponds**: pond created to reduce downstream sedimentation and expand waterfowl habitat;

- **Channel bank stabilization**: contour slopes to prevent slumping into streams, stabilization with rocks or gabion baskets at strategic locations, and enhance with suitable grass and shrub vegetation;

- **Buffer areas**: a transition zone between cropping/livestock fields and waterways designed to trap nutrients.[115]

The Drainage Act has undergone unfair criticism in some circles. Critics feel the act is too favourable to agriculture while ignoring wildlife interests. This project proved that fish habitat, water quality, and wildlife habitat can all be addressed through the Drainage Act. Further, some of the critics don't understand the importance of rural municipal drains for removing surplus water from towns and villages, thus preventing flooding around homes in the non-farming areas as well. Although there is the occasional disgruntled landowner who may resist an increase in their property assessment because of new or enhanced drainage requirements, the beauty of the Drainage Act is that local citizens are in charge, working through their municipality for the betterment of their communities.

---

115   Managing Agricultural Drains to Accommodate Wildlife. Brochure contained within final report 1993-1997. (OSCIA Archives).

American Chestnut Revival

OSCIA moved on to support other stewardship programs that gave landowners opportunities to participate in promoting biodiversity. One project in particular captured their imagination: saving the majestic American chestnut tree from the brink of extinction. OSCIA and its partners enthusiastically ventured into a program to plant seedlings and raise awareness of the American chestnut's plight. Farmers love their forests and willingly participate in tree planting, so there was little controversy for OSCIA to be supporting a program to bring back one of the most majestic trees ever.

Before the early 1900s, the American chestnut was an important tree species in eastern North America. It was referred to as the redwood of the eastern forests, a beautiful straight-grained deciduous hardwood that was resistant to rot. Today the American chestnut is nearly extinct due to a bark fungus imported more than a century ago on an Asian chestnut. Interestingly, a few American chestnut trees near Burford, Ontario, have not succumbed to the blight. Whether due to genetic resistance, luck, or environmental influences, it was motivation enough for the Grand River Conservation Authority to set up a tree nursery to propagate American chestnut tree seedlings from that grand old stock.[116]

---

116   Canadian Chestnut Council. Restoring the American Chestnut. An Endangered Species. www.canadianchestnutcouncil.ca/index.cfm?page=anEndangeredSpecies. (Accessed May 5, 2017).

Figure 8.1 The American Chestnut tree, profiled on OSCIA's 'Restoration of the American chestnut' project. (Source: Brochure in OSCIA archives)

In 1998, OSCIA launched a program entitled Restoration of the American Chestnut – Farmer Response to Species at Risk. OSCIA arranged for distribution of seedlings from the blight-resistant stock to two dozen farms across southern Ontario. Project manager Andrew Graham said, "The co-operator sites were chosen with a key objective of establishing American chestnut in areas of the province not yet affected by the blight. As an expected outcome, I recall most of the sites were in counties situated outside the traditional zone of American chestnut... The suitability of the seed source was another factor in selecting sites. Work performed by Forestry Canada at the time looked at characteristics of where the seed was originally harvested for the seedlings (i.e. Brant County, upper New York State) and aligned it for best fit with the candidate planting sites. The seed was harvested from mother trees that, at least until that time, had inexplicably escaped the blight. They were believed to have at least some blight tolerance. Much coaching was provided to OSCIA from experts with the Canadian Chestnut Council. Projects sites [included] the counties of Lanark, Huron, Oxford, Middlesex, Peterborough,

Northumberland, and Simcoe. Nothing could be more benign within the controversy over species at risk than encouraging new plantings of that once glorious tree admired by so many North Americans. The project was a success as the seedlings were snapped up quickly for suitable planting sites."[117]

Today, through the Canadian Chestnut Council, scientists from the University of Guelph and elsewhere in North America are working to identify blight-resistant varieties and expanding breeding programs aimed at re-establishing tolerant varieties of this majestic tree in the wild. Like the barn owl, our farm in Waterloo County is on the northern fringe of the historic range of the American chestnut. Will climate change, and the work of creative geneticists, open up new opportunities to restore the American chestnut for future generations to enjoy?

The success of the American chestnut project and the enthusiastic contribution of the farm community set the stage for OSCIA to expand its SAR efforts. The Eastern Loggerhead shrike caught our attention. In 2003, Wildlife Preservation Canada was invited to lead the recovery effort for this critically endangered songbird, which numbered fewer than thirty breeding pairs in a few isolated spots in southern Ontario.[118]

In Victoria County, the Nature Conservancy Canada (NCC) was active working with partners to protect the Carden Alvar, an outstanding example of rare alvar habitat. Alvars are characterized by a thin or non-existent layer of soil atop a limestone base that supports vegetation like cedar or juniper shrubs, grasses, and lichens. Located east of Lake Simcoe, the Carden Alvar totals 7,900 acres (3,200 hectares) of prime habitat for several bird species. Under the careful management of NCC and partners, the protected lands on the Carden Alvar will become a model for alvar ecology and stewardship in North America.[119]

---

117   Personal communication between Andrew Graham and Harold Rudy

118   Eastern Loggerhead Shrike. Wildlife Preservation Canada. www.wildlifepreservation.ca/loggerheadshrike/ (Accessed May 5, 2017).

119   Carden Alvar Natural Area. Nature Conservancy of Canada. "An Internationally Significant Area." www.natureconservancy.ca/en/where-we-work/ontario/our-work/delete/carden_alvar_natural_area.html (Accessed May 5, 2017).

OSCIA became engaged through promotion of the Loggerhead shrike recovery program. Farmers in cattle country from the Bruce Peninsula to the Ottawa Valley could have potential habitat. In May 2008, staff at the OSCIA Guelph office organized an outreach tour to the Carden Alvar. At that time, John Kinghorn, a cow-calf producer and OSCIA director from Victoria County, was passionate about the Loggerhead shrike project and engaged with beef farmers in the area to manage grazing lands for the encouragement of shrike habitat. The shallow soil of the Carden Alvar is primarily suitable for pasture land, leaving the shrubs and hawthorn trees that provide shelter and habitat suitable for birds like the Loggerhead shrike.

> All farmers and rural landowners appreciate wildlife, such as the beauty provided by the rare sighting of an owl, songbird, or amphibian. The "wild" part comes into play when certain populations are out of control

Visitors come from all over the world to the Carden Alvar, hoping to catch a glimpse of this rare songbird that's armed with the razor-sharp beak and talons to capture and impale their prey on thorns of hawthorn bushes.

During the staff tour to Loggerhead shrike country, we also learned about the competing interests at stake: the quarry industry coveted extraction of the high-quality aggregate found in the area, while conservationists sought to prevent the destruction of rare habitat. Mr. Kinghorn was a great mediator as the groups came together to find common solutions.

Establishing partnerships with conservation groups interested in protecting species at risk has been a good news story for OSCIA and good for farmers. Farmers do care and there is a genuine commitment to protecting habitat *and* their livelihoods. Financial incentives play a key role in attracting farmers' attention.

As part of its SAR outreach, OSCIA developed a booklet for children entitled *Species at Risk on Your Farm*. It provides a brief overview of the challenges facing the Milksnake, Fowler toad, American badger, Redside dace, Little brown bat, Bobolink, Monarch butterfly, Snapping turtle and

American chestnut. Complete with puzzles crafts and coloring opportunities, the booklet is interesting and educational for adults as well as children.[120]

Is it Wildlife or Wild Life?

There is a dark side to wildlife on the farm. Is "wildlife" one word or two? All farmers and rural landowners appreciate wildlife, such as the beauty provided by the rare sighting of an owl, songbird, or amphibian. The "wild" part comes into play when certain populations are out of control. When wildlife damages farm crops or kills livestock on a farmer's property, it affects the family's livelihood. And do we hear about it!

At almost every annual meeting of the OSCIA, some county or district puts forward a resolution asking the OSCIA provincial board to notify and consult with the appropriate government agency or wildlife group to find solutions to damage caused by one species or another. Canada geese, sandhill cranes, deer, black bears, wild turkey, raccoons, coyotes, and even songbirds in fruit crops can be a major frustration that threatens livelihoods. The challenge for OSCIA is to find the balance between promoting tree plantings, shelter belts, and hedgerows that support wildlife, and taking steps to control populations that may end up as nuisance species.

On the positive side, OMAFRA does provide compensation under the Ontario Wildlife Damage Compensation Program (OWDCP) to owners of livestock, poultry, and bees for damages caused by wildlife, typically from coyotes, wolves, and bears.[121]

For crop farmers, compensation for crop damage from wildlife is much more challenging. Crop insurance, administered in Ontario by Agricorp, does cover losses, as calculated within a farmer's overall yield averages. But spot losses, for example, along an acre or two of woodlot, are not covered.

---

120   Species at Risk on Your Farm. OSCIA. www.ontariosoilcrop.org/wp-content/uploads/2015/08/speciesatriskonyourfarm-web-rev.pdf. (Accessed May 5, 2017).

121   OMAFRA. Predation and Wildlife Damage. Ontario Wildlife Damage Compensation Program. www.omafra.gov.on.ca/english/livestock/predation.htm. (Accessed Aug. 21, 2017).

Sandhill cranes in northern Ontario have been known to destroy whole fields of crops, particularly in the spring after emergence.

Assessing Wildlife Impacts on Agriculture

In response to growing concerns over wildlife damage, OSCIA undertook research to get a better handle on the magnitude of the losses. In 1999, OSCIA developed a survey on behalf of the Ontario Agricultural Commodity Council to estimate the economic damage done by wildlife to cash crops, fruits, vegetables, and livestock. The survey sampled 1,000 farms in the spring and fall and 250 logbooks were distributed to growers so they could record wildlife damage with the help of Agricorp and OMAFRA field staff.

The original study was updated in 2009, and estimates at that time pegged the damage to Ontario commodities caused by wildlife at approximately $49 million per year.[122]

Dogs, Fences, and Wailers

Through various financial incentive programs since the year 2000, OSCIA has helped qualifying producers to mitigate wildlife damage. This funding was primarily provided by Agriculture and Agri-Food Canada through programs such as the Agricultural Policy Framework.

For a period, OSCIA provided cost-share funds to farmers who needed trained guard dogs such as Great Pyrenees to protect their sheep.

High fences to keep deer out of apple orchards are expensive, but also effective. Through various cost-share programs, financial assistance was provided by OSCIA and was appreciated by horticultural producers. OSCIA has delivered government incentives to provide netting to fruit and grape growers to prevent birds from feasting on their juicy crops. Ice wine grapes

---

122   Mussel, A. and C. Schmidt. "An Economic Impact of the Wildlife Impact Assessment for Ontario Agriculture." (2009) George Morris Center (OSCIA Archives).

are left exposed for months in the fall and early winter before they're ready for harvest, so netting is essential to keep the birds out.

The electronic age has provided "wailers" and "screamers" that can be programmed to sound like predator birds. To prevent complacency or familiarity with only one sound, multiple sounds of predators can be programmed in sequence to fool the pesky fruit eaters. These modern tools are much more socially acceptable to non-farming neighbours compared to the old propane bangers that sounded like shotguns or cannons being fired in the orchards every few minutes.

Most producers are reasonably tolerant of minimal or modest losses due to wildlife, but when a significant portion of their income is at stake, preventative steps need to be taken. OSCIA is frequently in discussions with farmers and government agencies about compensation strategies, whether it be through crop insurance, spot loss coverage, or public investment to seek solutions for landowners. Are they sufficient? In extreme circumstances, not likely. What will be the path forward for a more harmonious co-existence between wildlife and food producers?

Chapter 9

WHAT DO CROPS EAT FOR BREAKFAST?

Like most living organisms, plants absorb nutrients as food in order to survive. Farmers have known for millennia that all animals and humans excrete nutrients politely known as manure, along with other smelly gases. These manure nutrients can be recycled back into the food chain for healthier crops, and so the circle of life continues.

However, if manure is present, so too are bacteria and pathogens. Therefore, responsible manure management is essential. The Ontario Farm Environmental Coalition (OFEC) developed a Nutrient Management Strategy in 1998 to take an active role in establishing strategy and policy for nutrient management and water-quality protection. OFEC was indeed fortunate to have high-profile scientists including Dr. Gord Surgeoner, Dr. Gary Katchanoski, and Dr. John FitzGibbon to lead those efforts. Their expertise was the glue that bonded the many organizations together for a focused strategy.

By the late 1990s, numerous municipalities were moving to establish stricter rules that ranged from how and where new barns could be built, to how to manage field application of manure. Not every municipality viewed nutrient management in the same light; in some places, rules were minimal or non-existent. So it wasn't long before there was a hodgepodge of bylaws and standards that differed from one municipal boundary to the next.

By the time of the Walkerton Crisis in May 2000, OFEC had released a document titled *Nutrient Management Strategy* that included a draft bylaw for consideration by Ontario municipalities. But it soon became apparent that even those municipalities that adopted the bylaw were making substantial revisions. "As a result, Ontario ended up with the dreaded patchwork approach to nutrient management," according to David Armitage, who was secretary of OFEC in addition to his role at the Ontario Federation of Agriculture.[123]

> There was general consensus that the farming industry needed to become proactive to match crop requirements by using soil testing and manure analysis

With direction from OFEC and provincial government officials, it became apparent that one set of rules across the province was essential. In addition, "OFEC requested language indicating that regulations made under a provincial nutrient management statute supersede existing municipal bylaws addressing the same subject matter. This request was provided for in Section 61 of the Nutrient Management Act of 2002, and essentially addressed the municipal bylaw concerns."[124]

Regulating farm nutrients was a rather unusual situation where farm groups were actually asking for one provincial standard to be developed.[125] Regulations were drafted and implemented under the provincial Nutrient Management Act (2002) that made it mandatory for large livestock and poultry producers equivalent to three hundred Nutrient Units (NU)[126] or greater to develop a Nutrient Management Strategy (NMS) and a Nutrient Management Plan (NMP). The NMS would describe how they would match their manure production in general terms to a land base sufficient in size to accommodate the nutrients. For the NMP, each of the three hundred NU or larger farms would have to match site specific fields with manure incorporation to match the nutrient requirements for crops grown, as recommended by a government approved soil test.

---

123   Armitage, D., Director of Regulatory Reform, Ontario Federation of Agriculture. Personal communication with H. Rudy (May 9, 2017).
124   Ibid.
125   Graham, A. Executive Director, OSCIA. Personal communication with H. Rudy (May 12, 2017).
126   A nutrient unit (NU) is a measure of nitrogen and phosphorous contained in different types of manure. The calculation is used to compare the nutrient levels in manure generated by different types and sizes of farm operations.

There was consensus that the farming industry needed to become proactive to match crop requirements by using soil testing and manure analysis. Even purchased nutrients such as commercial fertilizer required more efficient alignment of soil test readings with crop needs. NMPs were a scientific way to better match manure and fertilizer nutrients to feed the crops. Soil testing to determine what nutrients are already available has been promoted by OSCIA and agri-business for decades.

Not all livestock and poultry farmers had this formula figured out, though. There was evidence back in 2000 that manure was being over applied, or applied at the wrong time. Winter spreading of manure was seen more as a disposal operation, as some farmers did not have appropriately sized manure storages to hold the contents until suitable soil and weather conditions prevailed during the growing season. By this time, Dr. John Fitzgibbon, professor at the University of Guelph's School of Environmental Design and Rural Development, stepped in as the chair of OFEC. His expertise on regulations and water quality was invaluable in establishing new policies.

In response to requests from farm leaders wanting to learn how farmers in other jurisdictions were responding to manure application challenges, a trip to Quebec was organized in November 1998 by OSCIA and OFA's Tiffany Svenssen, who was also the project co-ordinator for the Water Quality Working Group. We heard nutrient management plans were becoming mandatory in Quebec and, as in New York State, there were generous grants to assist producers in storing and managing their manure. The Quebec Environmental Farm Plan was structured with a strong focus on nutrient management involving local clubs, called Clubs-conseils, each made up of about twenty farmers who worked with an agronomist to design and implement their nutrient management requirements. The club concept began in 1993 and was seen as a highly successful way to engage producers.[127]

---

127   Foulds, C. "The farmer club approach in Quebec helps transition to sustainable agriculture." Sustainable Farming (McGill University) www.eap.mcgill.ca/MagRack/SF/Summer%2095%20B.htm (Accessed Aug. 25, 2017).

The Quebec trip involved two passenger vans of about ten passengers each from various provincial agencies and organizations. Tiffany drove one van and I drove the other for the agenda-packed three-day visit. The stops included farm visits to hear about how farmers were complying with their new nutrient management regulations, agri-environmental farmer club participants, government staff extension efforts, a municipal perspective from the city of St. Hyacinth, and a final stop on the return trip at the Water and Earth Sciences Associates facility in Kingston, Ont.

It was clear that financial assistance was required to transition livestock and poultry farmers to meet new regulations under the Ontario Nutrient Management Act. The regulations required significant changes for some producers. No more winter spreading, no more high-trajectory manure guns shooting brown clouds of precipitation (and odour) into the air. Farm leaders lobbied long and hard for financial assistance. In October 2004, the government announced the $20 million Nutrient Management Financial Assistance Program (NMFAP), and OSCIA entered into an agreement with OMAFRA to deliver it.[128]

Scientists and government experts recommended that the most effective way to target the program was to start with producers who managed large numbers of livestock. The farm operation's size would be measured using the NU calculations as defined under the protocols contained in the act.[129] Each farm would have to generate three hundred or more NU to be subject to the regulations and to be eligible for funding under NMFAP.

Demand was so strong that OSCIA was short of funds needed to help bring all the eligible producers into compliance. In the end, we believe that half of the large livestock operations participated. The program was extended until the fall of 2006 with another $3.7 million provided by the

---

128 "Government helps farmers to better protect the environment." News release. Government of Ontario, Ministry of Agriculture, Food and Rural Affairs (October 14, 2004) (Accessed May 8, 2017).

129 June 2003 Nutrient Management Protocol for Ontario Regulation 267/03 Made Under the Nutrient Management Act, 2002. Ontario Ministry of Agriculture, Food and Rural Affairs. www.omafra.gov.on.ca/english/nm/regs/nmpro/nmpro03-jun03.htm (Accessed Feb. 14, 2017).

province to wrap up this successful transition for the largest livestock and poultry producers.[130]

Given its successful track record delivering cost-share programs in support of on-farm environmental improvements, it was a natural fit for OSCIA to deliver NMFAP. What was not natural, though, was for OSCIA to invite owners of large livestock and poultry operations to public meetings about a new program targeting the largest producers, many of them with multiple farms. As shrewd business people, about half of the 7,300 large NU farms made every effort to maximize their financial cost-share. The guidelines for NMFAP were complicated and their questions detailed and probing. NMFAP, when combined with the federal Agricultural Policy Framework funding introduced in 2005, offered a potential of up to $100,000 in cost share per farm unit.[131] This got everyone's attention. A farm business could have more than one Farm Unit for application purposes. Keeping track of eligibility to ensure the program was fair and equitable presented additional challenges.

With all its complexity and challenges, NMFAP had the potential to bring OSCIA to its knees, according to one of OSCIA directors. Fortunately, it did not. The short time frame for delivery required project completion by farmers and inspection by our county/district field staff by December 31, 2005, and then to process all those claims in the Guelph office and mail out the cheques. Our office had to close the books by March 31, 2006.

Staff did a superb job of explaining the rules, reviewing applications, and inspecting claims. I took personal ownership of verifying and double-checking the complicated family ownerships of multiple farm units. In my personal file cabinet, I kept a ¾-inch-thick wad of computer printouts with my hand-written notes from discussions with these producers about their particular circumstances. I got to know the business structure of the larger operations quite well—at least on paper.

---

130 "Ontario government provides additional funding for nutrient management." News release. Government of Ontario, Ministry of Agriculture, Food and Rural Affairs. (Accessed May 5, 2017).

131 A farm unit can be one parcel of land under a single deed; or all parcels of land owned by an individual, partnership or corporation; or combination of different properties.

It was a challenge for OSCIA's program delivery infrastructure to implement over $23.7 million in cost-share support under such tight deadlines. Our systems were stretched by the need to program a database to calculate the nuances of eligibility criteria and record approvals and claims. To hit the ground running, we used an Excel spreadsheet to track the first flush of paperwork. Various lists had to be checked and scrutinized carefully to ensure compliance. Eventually, the individual Excel lists were integrated into the master data base to ultimately balance the books. Indeed, OSCIA may have been brought to one knee during this process, but we got through it unscathed and were soon back on our feet, having met all program and audit requirements. The NMFAP was a sellout.

> Through the use of soil management techniques like no-till planting, cover crops, and the addition of organic matter, farmers can boost the health of their soil and help mitigate climate change

More on Crop Diets - The Invisible Food

By the time they're in high school, most students are familiar with the concept of photosynthesis: carbon dioxide + water + sunlight = sucrose (biomass) + oxygen. Before the end of primary school, they're introduced to the important role played by plants in balancing the global carbon budget when they absorb sunlight for energy and remove carbon dioxide, a potent greenhouse gas (GHG).[132]

Agricultural plants require carbon dioxide, some of which goes back into the soil as organic matter to absorb moisture and convert nutrients into food to nourish living plants. OSCIA has always recognized the benefits of managing soils and crops to maximize the amount of organic matter and the role of healthy soils in sequestering carbon dioxide. These chemical processes are essential to our survival!

---

132   Nature's Sweet Mystery: An Integrated Curriculum Resource for Grades 4-6. www.sugar.ca/SUGAR/media/Sugar-Main/News/Natures_Sweet_mystery.pdf (Accessed May 8, 2017).

I started attending meetings in the mid-1990s to learn more about the role of agriculture in reducing greenhouse gases that affect climate change. The Soil Conservation Council of Canada (SCCC), of which OSCIA is a member, has been a leader in raising awareness and influencing how agriculture can contribute to reducing the amount of carbon released into the atmosphere. Specifically, how can we use plants to store carbon as organic matter in the soil and, in effect, turn our farms into carbon sinks? Through the use of soil management techniques like no-till planting, cover crops, and the addition of organic matter, farmers can boost the health of their soil and help mitigate climate change. How much would organic matter increase by improvements to crop residue management? Scientists across the country were delving into this research.

Furthermore, could other improvements on the farm help offset what industrial factories were expelling? Other greenhouse gases such as nitrous oxide and methane are also by-products of agricultural production.[133] The SCCC, led by farm leaders in Western Canada, was in talks with electrical utilities to discuss emissions trading scenarios. Utility companies power our cities and factories but they are also large emitters of carbon. By paying offset fees to other industries such as agriculture, they could compensate for their emissions until years down the road when they could afford to replace older equipment with new less-emitting technology. This way, everyone would win! But it is not that simple.

Although carbon sequestration discussions have been underway since the 1990s, we still do not have a credible carbon trading system for agriculture. We may be close. The Ontario Ministry of Environment and Climate Change is committed to establishing a cap and trade system for Ontario.[134] By the spring of 2018, Ontario had introduced a proposed Voluntary Carbon Offset program framework.[135]

---

133   Overview of Greenhouse Gases. United States Environmental Protection Agency. www.epa.gov/ghgemissions/overview-greenhouse-gases. (Accessed May 8, 2017).

134   Cap and Trade. Government of Ontario. www.ontario.ca/page/cap-and-trade (Accessed May 8, 2017).

135   Ontario's Carbon Offset Programs, https://www.ontario.ca/page/ontarios-carbon-offsets-programs, (Accessed May 8, 2018)

It takes a long time to increase the amount of organic matter in soil by one per cent. Research has shown that plant roots may actually be more effective at creating and retaining carbon than the surface residue left by no-till management. According to one study, "By the end of the season, 52 per cent of the root-derived C was still present in the soil, whereas only four per cent of (surface) residue-derived C remained, thereby showing clear differences in the fates of root and residue-derived C."[136]

Ontario research has shown no-till residue management on its own produces little overall carbon accumulation.[137] However, where a cereal crop such as wheat is added to a crop rotation of corn and soybeans, soil organic carbon can be increased by up to 14 per cent compared to conventional tillage.[138] Manure, biosolids, and compost along with annual cover crops will also significantly increase organic matter and boost carbon sequestration over time.[139]

The book is still being written about the complexities of science, and the opportunities and threats for farmers under a cap and trade system. OSCIA has many producers working in co-operation with OMAFRA field staff on cover crops and monitoring results for organic matter improvement and soil health. OSCIA has not taken a proactive role to develop protocols to meet

---

136 Kong and Six. "Tracing Root vs. Residue Carbon into Soils from Conventional and Alternative Cropping Systems." Journal of Soil Biology and Biochemistry (July-August 2010), p. 1209.

137 Murage, E.W., P.R. Voroney and R.P. Bayaert. "Dynamics and Turnover of Soil Organic Matter as Affected by Tillage." www.uoguelph.ca/plant/research/agronomy/publications/pdf/murage%20et%20al.pdf (Accessed May 8, 2017).

138 Van Eerd, L.L., K.A. Congreves, A. Hayes, A. Verhallen, and D.C. Hooker. "Long-term tillage and soil crop rotation effects on soil quality, organic carbon, and total nitrogen." Canadian Journal of Soil Science Vol. 96 p. 303-315 (Accessed May 8, 2017).

139 Fronning, B.E., K.D. Thelen and D.H. Min. "Use of Manure, Compost and Cover Crops to Supplant Crop Residue Carbon in Corn Stover Removed Cropping Systems." ResearchGate Publications (Nov. 2008). www.researchgate.net/publication/216811315_Use_of_Manure_Compost_and_Cover_Crops_to_Supplant_Crop_Residue_Carbon_in_Corn_Stover_Removed_Cropping_Systems (Accessed May 8, 2017).

carbon trading requirements—not yet at least. A common fear expressed at farm meetings that I have attended is that the cost involved in verifying carbon sequestration may exceed the value obtained under a cap and trade system. Oh, and what about those leading-edge farmers and superior soil management innovators who have been capturing carbon over the past twenty-five years? Will they get retroactive credits? Not likely, but by their healthier soils by now should be padding their pocket books with improved productivity.

> But alfalfa is the queen of legumes because it combines superior soil-building characteristics with plentiful high-quality protein for livestock while it also captures nitrogen from the air

I remain optimistic that somewhere, somehow, sometime, there will be new opportunities for farmers to receive credits for superior soil management techniques. At a minimum, we all benefit from higher crop productivity, more resilience to drought and environmental improvements that result from their efforts. At most, farmers may earn a few more bucks per acre, perhaps enough to take the family on a vacation each year!

Free Nutrients: The Ontario Forage Masters Program

Why not let the plants harvest their own food and deposit nutrients into the soil? With legumes, that is in fact the case. Legumes have the unique ability to form a soil partnership with *rhizobia* bacteria that facilitate the formation of nodules which capture nitrogen from the air and make it available to the plants while also improving the soil's structure.[140] Legumes include alfalfa, beans (e.g. soya, pinto, navy), peas, lentils, and peanuts. Forage legumes such as red clover, birdsfoot trefoil, and sweet clover also have N-fixing and soil building characteristics, and may be used for livestock feed or cover crops.

---

140   Soil Improvements With Legumes. Government of Saskatchewan. www.saskatchewan.ca/business/agriculture-natural-resources-and-industry/agribusiness-farmers-and-ranchers/crops-and-irrigation/soils-fertility-and-nutrients/soil-improvements-with-legumes. (Accessed Sept. 25, 2017).

Alfalfa is the queen of legumes because it combines superior soil-building characteristics with plentiful high-quality protein for livestock while it also captures nitrogen from the air. Harvested as dry hay or higher moisture silage, alfalfa (commonly referred to as "forage") may be planted with other grasses such as timothy to diversify the feed ration and fill in the field gaps if alfalfa winterkills.

Forage Masters

In an ideal world, every farm would grow alfalfa in rotation with other crops, but what would we do with all that hay? So far at least, humans don't eat hay. With all its exceptional characteristics, does alfalfa receive the recognition it deserves? Is there a National Alfalfa Day, or an International Year of Alfalfa? Not yet, it seems. In 1987, however, OSCIA established the Ontario Forage Masters Program to bring recognition to the value of hay and pasture crops.

Organized by OSCIA's then secretary-manager, Doug Wagner, the Ontario Forage Masters Program was designed to:

- Identify forage producers with winning management practices from among local soil and crop improvement associations, and

- Recognize outstanding forage producers and promote excellence in growing, harvesting, and storage of forages.[141]

Doug credits three individuals for their creative thinking to create the Ontario Forage Masters Program. John Benham, a strong advocate and experienced dairy farmer, knew the value of forages from his personal experiences. Dr. Stan Young, extension co-ordinator and professor in the crop science department at the University of Guelph, was an enthusiastic supporter, and John Lawrence, general manager at Dekalb Seeds, provided expertise on the scoring system to identify quality forages. Doug looked after the promotion to the counties/districts for OSCIA.

---

141   Ontario Forage Masters Program Guidelines. OSCIA internal files.

What Do Crops Eat for Breakfast?

The challenge was issued to each OSCIA county/district across the province to enroll farmers in forage competitions in their area. Judges would visit each farm entrant to assess their agronomic management and obtain a forage sample that was sent to a laboratory for analysis. Samples were analyzed for crude protein, acid detergent fibre (ADF), and neutral detergent fibre (NDF)[142] to determine the feed value score. See the scorecard below:

WEIGHTING OF JUDGING CRITERIA FOR TOTAL SCORE

All stored feed samples obtained by the judges will be analyzed for crude protein, ADF, NDF, with equal weighting to determine the feed value score.

SCORE CARD	WEIGHTING (%)
AGRONOMIC FACTORS SCORE CARD	50
HARVEST AND STORAGE FACTORS SCORE CARD	25
FEED VALUE SCORE	25
	100%

Figure 9.1 Forage Masters Program Scorecard (Source: OSCIA archives)

ADF and NDF are values used in laboratory feed analysis to measure forage quality according to digestibility, energy and voluntary fiber intake. (Source: OSCIA)

First-place winners from each county/district would then be invited to prepare a presentation to be given in the provincial competition held each year at the Royal Agricultural Winter Fair in Toronto. Each year's Ontario Forage Masters winner was also invited to attend and give their presentation at the American Forage and Grasslands Conference held annually at various locations across the United States.

---

142 OMAFRA Factsheet, "Feed Analysis Explained," : http://www.omafra.gov.on.ca/english/livestock/dairy/facts/16-049.htm (Accessed Sept. 27, 2017).

Figure 9.2 The 2007 Awards Luncheon at the Royal Agricultural Winter Fair. Perth county's Maplevue Farms was recognized as the 2007 Ontario Forage Master award winner. Left to right: Paul Wight, Pickseed Canada (sponsor), Laura and Doug Johnston (Maplevue Farms), Frank Hoftyzer (2007 OSCIA President), Ron Piett, Agri-Food Laboratories (sponsor). (Photo Source: OSCIA)

The design of the Ontario Forage Masters Program has evolved over time. It is quite costly to support provincial competitions and provide travel expenses to the Royal and the American Forage and Grassland Conference each year. It would not have happened were it not for the initial generous sponsors including NK Seeds, Dekalb, Novartis, International Stock Foods, Agri-Foods Laboratory, and Syngenta. More recently, Pickseed Canada, SGS Canada, and the Royal Agriculture Winter Fair stepped up with their support. Laboratory analysis was generously provided at no charge by SGS Laboratories. Additionally, Paul Wight, Pickseed Canada's sales manager for Ontario and Atlantic Canada, volunteered his time to tally the final scorecards. The names of the annual Ontario Forage Masters are listed in Appendix 7.

What Do Crops Eat for Breakfast?

Alfalfa is an important crop that captures nitrogen, an essential nutrient that contributes to better yields in subsequent crops and helps reduce or eliminate the need for purchases of commercial fertilizer. Alfalfa and other forage grasses play a vital role in crop rotation to help ensure healthy soils for generations to come.

Today, the Ontario Forage Masters Program has evolved into a self-assessment process to reduce the reliance on local volunteers. The annual cost of supporting this thirty-year old program has also been reduced. The program continues to play an important part in promoting the value of forages and the role of Ontario's hay and pasturelands as soil protectors and soil fixers.

Chapter 10

OSCIA's Evolving Role under Federal/Provincial/Territorial Frameworks

Being entrusted with the distribution of cost-share funds to farmers had some unique challenges. OSCIA field staff worked mostly out of their home offices, and in many situations, on their farm properties. Farmers don't punch the clock at 8 a.m., nor at 5 p.m., so neither can we. The work day extends throughout the waking hours. Phone calls, application drop-offs, or pop-ins for clarification on program eligibility have been known to happen before sunrise and long after sunset. Field staff have been incredibly accommodating to clients in OSCIA's era of program delivery.

> Those who showed up at 8 a.m. anticipating prompt service were so far back in the line that funds were already committed

When funds are limited, and pent-up demand bubbles over in anticipation of cost-share opportunities like those associated with the EFP program, lineups begin early. We've had examples where those in the queue at 5 a.m. were successful in getting through the doors when they opened at 8 a.m., but those who showed up at 8 a.m. anticipating prompt service were so far back in the line that funds were already committed by the time they got to the door. This scenario played out numerous times across Ontario during the Growing Forward program.

Over the years, there was considerable evolution in program design as we tried to find the right mix of eligibility requirements and cost-share support in order make optimal use of limited public funds while managing stakeholder expectations.

Lacking in those early years was a method of measuring and grading projects for their ability to improve environmental conditions. How many tonnes of topsoil would be saved by approving a no-till drill? How much could phosphorus in waterways be reduced by helping farmers build larger manure storage facilities to ensure application was timelier? Similarly, by how much could we improve biodiversity—and how would we measure it—if we approved a tree planting project, or assisted a landowner with a native grass plantation? The evolution of program design dictated that we needed more rigorous metrics to describe environmental impacts. Then we could target funds for best use. After all, it was the public purse supporting our efforts.

Through the early years of the Agriculture Policy Framework (2003-2008), we began to investigate options for prioritizing projects, such as the Environmental Benefits Index (EBI) introduced by the U.S Department of Agriculture to score and identify marginal farmland to be removed from agricultural production.[143]

I included a discussion on the merits of an EBI ranking system in the final paper of my master's program. We were confident that a similar formula could provide scoring criteria to assist in scoring projects for approval. Factors such as slope of land, soil type, or proximity to a watercourse could help us.[144]

For the National Soil Conservation Program, discussed earlier, we indeed did have a formula for our county/district review committees to rank applications. A priority system had, to some degree, set a precedent and was successful. Close to half the projects were declined or, conversely, those projects that demonstrated the best value were approved.

---

143   USDA (1999) Environmental Benefits Index. Factsheet on Conservation Reserve Program Signup 20. Farm Service Agency, Washington, D.C.

144   Rudy, H. (August 2003). "Performance Measures for Environmental Programs using the Environmental Farm Plan as the Basis for Analysis" (University of Guelph).

Leaner Financial Years

The mid- to late-1990s were leaner financial years that forced OSCIA and partners to look at alternative cost-share approval processes that did a better job of targeting and focusing upon specific environmental objectives. The more generous opportunities of the Land Stewardship and Permanent Cover programs, which ended in the early 1990s, would not likely return in the new era of tighter budgets. The Green Plan supported a long list of projects from 1992 to 1997, including the EFP, but on-farm cost-share incentives were reduced.[145]

After 1997, we expanded our network of collaborators until the first federal/provincial/territorial Agricultural Policy Framework (AFP) came about in 2003 (it was actually 2005 when funds from the AFP began to flow). Funding was required to support EFP workshops for producers, conduct peer review of their action plans, provide modest cost-share support for improvements, and cover general administration.

The Ontario Farm Environmental Coalition (OFEC) continued its strong moral and verbal support for EFP. OFEC was not a legal entity so could not apply for and manage funds. However, the Ontario Federation of Agriculture (OFA) provided extensive leadership for OFEC, through individuals like David Armitage, OFA's senior policy adviser at the time. OFA, on behalf of OFEC, submitted an application to the Ontario Agricultural Adaptation Council (AAC) to support EFP. This strong coalition of support encouraged the continued funding of the EFP after the Canada-Ontario Green Plan ended in 1997. For example, the AAC provided $8.5 million over the next three years, then approximately $8 million more over two agreements up to 2004. The OFA was a strong advocate for EFP, transferring the funding to support OSCIA for EFP workshops, training, communication, and cost-share for farmers. In addition to providing funds for EFP, the AAC supported 25 separate OSCIA projects over the years with an investment of over $1 million. (See Appendix 8 for a complete list.)

---

145   Canada-Ontario Environmental Green Plan. www.agrienvarchive.ca/gp/gphompag.html. (Accessed Sept. 7, 2017).

Canada-Ontario Agricultural Policy Framework

The Agricultural Policy Framework (APF) agreements officially got underway in 2005 to support the EFP. Cost-share opportunities to implement beneficial management practices (BMPs) consisted of the Canada-Ontario Farm Stewardship Program (COFSP), Green Cover Canada Program, and the Canada-Ontario Water Supply Expansion Program. COFSP funding was channeled through the OFA on behalf of OFEC. A combined list of eligible projects under COFSP, Greencover Canada, and the Canada-Ontario Water Supply Expansion Program is outlined in Appendix 9.

OSCIA received close to $60 million through the APF. By November 2007, 729 EFP workshops had been held with over 11,000 participants enrolled in the APF generation of programs.

Figure 10.1 - Outlines the Top 12 On-Farm Improvements out of the potential best practices categories (November 2007). For example, Category 14 'Improved Cropping Systems', identifies 2,372 projects completed. (Source: OSCIA internal archives)

Growing Forward

Both EFP and COFSP continued under Growing Forward (2008-2013), which was supported by a federal/provincial/territorial framework and funded by AAFC and OMAFRA.

> "We needed the ability to direct funding to projects that had the greatest potential to satisfy environmental priorities."
> (Andrew Graham, OSCIA executive director)

Figure 10.2 A peak month in 2007 with six hundred cheques having cleared the bank and waiting for reconciliation. (Photo Source: OSCIA)

OSCIA's experiences with delivering Growing Forward exposed the vulnerabilities of first-come, first served program models, according to Andrew Graham, executive director.

"The program had insufficient funds to satisfy heightened demands from a farm community that had grown to expect program funding levels that paralleled the previous APF. Many producers had begrudgingly accepted what needed to be done to try and receive an allocation. We made the best of a less-than-perfect situation, but the signals were clear that program design needed to evolve. We needed the ability to direct funding to projects that had the greatest potential to satisfy environmental priorities (i.e. make the biggest difference).

"We (OSCIA) secured funding through AAFC to conduct a literature search and develop a framework for a riparian program that would utilize a competitive bid model. It called for an Environmental Benefit Index (EBI) to help isolate top projects that best met program objectives.

"OSCIA got the chance a short time later to apply the competitive bid concept to [secure] agricultural grasslands to benefit Bobolinks. An EBI was developed by OSCIA with assistance from noted bird biologists to ensure the science was correct and the corresponding rankings were defendable. The program was initially supported by MNR, with Environment and Climate Change Canada coming on board the second year with additional funds to include the Eastern meadowlark.

"The experience and success gained through the Grassland Stewardship Program led to more opportunities to apply the competitive bid design. OMAFRA provided funds for a program to promote BMPs in riparian areas of Lake Simcoe watershed a few years back, and OSCIA continues with a grassland initiative funded through Species at Risk Partnerships on Agricultural Land (SARPAL)."[146]

SARPAL is now broadened to cover a wider range of habitat encompassing grassland stewardship, grassland birds, and habitat for the American badger, an endangered species in Canada.[147]

Program administrators struggled to find a more equitable application intake. Exclusive online application intakes were tried too, but like acquiring concert tickets to see a famous rock star, the new opportunity was often sold out in a few minutes.

By 2008, we had enrolled close to 40,000 farm businesses through the EFP education and training program, but the available cost-share funds did not come close to satisfying the demand.

There had to be a better way. First-come, first-served had been our application approval model since the days of the Land Stewardship Program.

---

146 Graham, A. OSCIA executive director. Personal communication with H. Rudy. (Oct. 19, 2017).

147 SARPAL. Species at Risk Partnership on Agricultural Lands. www.ontariosoilcrop.org/oscia-programs/sarpal (Accessed Oct. 20, 2017).

Eligibility requirements were quite basic. At that time, first-come, first-served worked well, as enrollment was slower, more cautious, and methodical, resulting in funds being dispersed through most of the program cycle.

Our pilot project, with a made-in-Ontario EBI, focused on the Bobolink and Meadowlark, was successful. As we reached the end of the Growing Forward program in 2013, policy-makers were pulling out all the stops to conceive a program design model that would incorporate a merit-based system that encouraged applicants to focus on justification for their priorities. Growing Forward 2 marked a dramatic shift in application procedures and approval processes.

Growing Forward 2

> *"What scared me the most was Harold sliding the Growing Forward 2 agreement between OSCIA and OMAFRA in front of me, saying it was the biggest dollar agreement to date for the association and I needed to sign it."*
>
> — Henry Denotter, 2013 OSCIA President (Essex County)

Growing Forward 2 (GF2), launched in 2013, was designed to encourage innovation, competitiveness, and market development in Canada's agri-food and agri-products sector through cost-share funding opportunities. Overall, the program has provided over $142 million in cost-share funding for more than 6,100 projects in Ontario since 2013.

First-come, first-served was replaced by intake dates, after which scoring criteria were applied in a merit-based approval process. Webinars provided instructions for program expectations and how to complete a strong application. Six focus areas were eligible for funding:

- Environment and climate change adaptation
- Assurance systems (food safety, traceability, animal welfare)
- Market development
- Animal and plant health

- Labour productivity enhancement
- Business and leadership development[148]

To meet the program's merit-based evaluation criteria, a project had to address a demonstrated need or business improvement as identified through an action plan from the EFP, Growing Your Farm Profits, or food safety self-assessment. Eligible projects could receive up to 50 per cent cost-share funding for planning-type projects and 35 per cent for implementation. But only the very best project would be funded. The merit-based approval process required additional administration and scrutiny, but overall it has hit the target with quality applications.

How Did Farmers Adjust to Merit-Based Approval?

For twenty-five years, applying for cost-share funding for environmental projects was easy. Farmers would provide basic tombstone information (name, address, phone number, etc.) and identify a need for the project. Sometimes proof of a building permit or an engineering design or some other approval was required, depending on the project. Field staff would often meet one-on-one with producers to help them complete the application.

The rules for merit-based applications were a game-changer.

Internally, OSCIA responded by restructuring administrative staff with expertise in three categories: business planning, food safety/biosecurity, and environment. These three units became a great resource for field staff and applicants. Responsibility for reviewing and understanding the GF2 program guide lay strictly on the shoulders of producers. Staff could coach a producer to ensure they understood the guide but did not assist in completing each project proposal.

Initially, some were surprised by the shift in the approval process; a high number of merit-based proposals were declined in the early days, but after

---

148 Growing Forward 2. Getting Started. OSCIA Programs. http://www.ontariosoilcrop.org/oscia-programs/growing-forward-2/growing-forward-2-step-by-step/ (Accessed Oct. 24, 2017).

a few intakes, the quality of applications improved dramatically. Farmers had more time to do their homework, acquire permit approvals, and consult experts before the intake dates to ensure their projects were high quality.

Merit-based approvals increased the certainty that the ministry's priorities were being funded. Biosecurity and food safety were new topics for OSCIA but by tapping outside expertise when needed—for example, consulting veterinarians about biosecurity issues—the merit-based approach continues to pay dividends by providing high-quality projects.

How have Farmers and Society Benefited from GF2?

Here's how OMAFRA has graphically illustrated our collective accomplishments for GF2:[149]

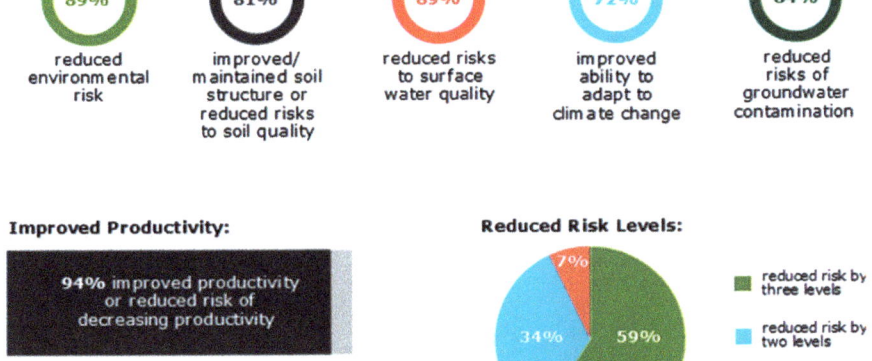

Figure 10.3 Growing Forward 2 for producers. Spotlight on Environment and Climate Change, December, 2016. (Source: OMAFRA)

---

149   Growing Forward 2 for Producers. Spotlight on Environment and Climate Change. OMAFRA. http://www.omafra.gov.on.ca/english/about/growingforward/gf2-farmbus.htm. (Accessed Oct. 26, 2017).

Growing Forward 2 has helped producers improve soil quality, reduce environmental impacts and improve adaptability to climate change. As with previous programs, GF2 has been a sellout.

Next Policy Framework - Canadian Agricultural Partnership

Officially announced in February 2018, the wheels are in motion to establish the next agricultural policy framework to take us to 2023. This next generation is to be called the Canadian Agricultural Partnership. Senior government leaders and politicians have announced that we are advanced in our discussions for a smooth transition from GF2. That is great news for the Canadian agriculture and food industry.

What will the Canadian Agricultural Partnership look like? Here's a summary communicated by AAFC on their website in February 2018:

The Canadian Agricultural Partnership will help the sector:

- "Grow trade and expand markets to seize key opportunities and address emerging needs;

- Advance science and innovation, with an emphasis on innovation and sustainable growth; and

- Better reflect the diversity of our communities, enhance collaboration across different jurisdictions and secure and support public trust."[150]

What opportunities will emerge for OSCIA as well as agricultural and food producers under the Canadian Agricultural Partnership? We are encouraged by continuing accolades for the EFP and support for a national EFP that establishes minimum standards. Additionally, progress with the Sustainable Farm and Food Initiative is a golden opportunity to brand the Canadian agri-food sector to meet or exceed any international market requirements for being sustainably sourced. It can also be the stimulus for farms to implement

---

150 Agriculture and Agri-Food Canada. Canadian Agricultural Partnership. http://www.agr.gc.ca/resources/prod/doc/cap/cap_factsheet_feb18-eng.pdf (Accessed May 8, 2018)

a broader range of best practices to address climate change. OSCIA staff have been exceptional at administering the many components of GF2, so why wouldn't we build that success under the Canadian Agricultural Partnership?

Chapter 11
Modus Operandi

Thirty-Two Presidents as Boss

"You're sleeping in my bed," uttered the gravelly voice, as I awakened to a 5:30 a.m. sunrise. Was this some real-life version of *Goldilocks and the Three Bears*? I wondered as the ornate door to my bedroom was flung open. I was a guest at a luxurious B & B, where I stayed occasionally after evening meetings in Guelph. My colleague Denis Perrault, OSCIA's 1998 president, was the one who introduced me to a stately bed and breakfast at reasonable rates. Denis and I had arrived late the previous evening while the innkeeper was away for the evening. With no innkeeper to direct us, I was quite sure that I was to take the room on the left at the top of the stairs, while Denis was in the bedroom to the right.

We drank a shot of brandy (complementary for guests in each room) as we reflected on the day's business, and then turned in. I should have been suspicious about my designated bedroom when I saw the single garment bag and narrow briefcase in the huge closet—but I never gave it a thought. It was unsettling to be awakened at dawn by a stranger invading my unlocked room and declaring that I was sleeping in his bed. From the sound of his insistent French accent and mostly unintelligible words, less than fresh appearance, and nightclub aromas that followed him into the room, I assumed he had walked right in off the street! I wasn't about to argue with him and started to gather my possessions, first grabbing my wallet and car keys. Denis heard the commotion and appeared at my doorway. My lack of comprehension to what was transpiring was directly related to the fact that my French language skills were particularly deficient at 5:30 a.m. Thankfully, Denis, who promptly mediated in French, concluded that I was, in fact, sleeping in this stranger's bed and that the garment bag and briefcase in the closet were his. My room should have been to the left at the bottom of the stairs, not the top of the stairs. Needless to say, there was no more sleeping that morning. I hastily exited with my possessions, got dressed, and then lounged in a chair on the veranda as the sun came up on a fine summer morning.

> The mentoring process allows the VPs to be fully engaged in the discussions and decisions

Sleeping arrangements were not always so memorable, but the thirty-plus OSCIA presidents whom I served were outstanding citizens, a pleasure to work with and all dedicated to betterment of the agricultural industry and their communities. They had a keen awareness of environmental impacts and were always striving for soil and crop improvement. I often wonder whether anyone else can lay claim to working full-time under the direction of over thirty presidents. Would this be a world record? I typed in "employment time" in the search window on the Guinness World Record website and turned up 12,516 results! Perhaps this quest for knowledge will be a retirement project.

The OSCIA governance structure permitted an election process that encouraged a new president each year. The OSCIA constitution didn't require it, but a tradition had evolved whereby the first vice-president moved up to president each year, and the ruling president remained on the board and executive committee as past-president. In similar fashion, the second vice-president stepped into the first vice-president's position, and the third moved up to second. All these positions went through a nomination and election process but at no time during my tenure, did the anticipated outcome of these elections waver. About the only question in doubt at election time was which person on the board of directors would fill the vacant third vice-president's position. As Barry Hill, 2010 president, so aptly pointed out, the process was in effect a three-year mentorship where politics is cast aside to groom each year's new president. In 2008, we modified the OSCIA constitution so that the incoming president would be elected at the summer meeting, then serve as president-elect until ratification by delegates at the annual meeting in February. The balance of the executive committee and board members were also ratified at this time. Our government partners liked this model because it provided a predictable, smooth transition from one president to the next. It was also a comfort to our close funding partners, who were relying on OSCIA to deliver multimillion-dollar agreements.

> *"In 2009, we continued to focus on raising the profile of OSCIA. This was accomplished by working with some of the best farmers and with the leadership and initiative of the local soil and crop clubs and provincial staff."*
>
> *-Pat Lee, 2008 OSCIA President (Oxford County)*

As senior manager, I didn't mind working with a new boss each year, although it was challenging to let go of a fruitful working relationship after establishing a great rapport with these fine leaders during their year of fame. Accommodating a new style each year did require some adjustment. Some presidents were very much hands-off, as they had busy and involved farm operations. I recall 2003 president Lloyd Crowe asking me about the time commitment required as he considered serving on the executive committee. I reassured him that when he became president, he could delegate duties to other executive members or directors so that his farm business would not be neglected. My adjustment to a new president each year, however, was trivial in comparison to the adjustment that would be required in a politically volatile organization, or perhaps worse, one in which the same old guard kept getting re-elected each year. Under our constitution, no director could serve on the board for longer than ten years (including moving through the executive committee). There was never complacency or dead wood at OSCIA. The board had just enough turnover each year to keep the meetings fresh and dynamic.

Early in my career at OSCIA, I questioned the logic of electing a president for just one year. Although the constitution and the electoral process allows a president to be re-elected for a second term, it has never happened. In total, by the time I retire in 2018, I will have worked with thirty-two presidents since joining OSCIA in 1987. All of them were gems to work with. Now I see the process of mentorship and predictability as a distinct advantage. The mentoring process allows the VPs to be fully engaged in the discussions and decisions. Presidents at OSCIA have never been on an ego trip. In fact, the immediate past-president remains fully engaged, contributing wisdom and experience to help row the boat. It is a team effort and a solid governance model that has served the Ontario agricultural community well.

> OSCIA's current strategic plan identifies gender imbalance as a weakness

Paddling in the right direction

Speaking of rowing the boat, another tradition was launched in 2007 when then-president Frank Hoftyzer was presented with a wooden canoe paddle in appreciation of OSCIA service. The paddle was presented by Ginty Jocius, president of Canada's Outdoor Farm Show. Pat Lee, the first vice-president, was in charge of hosting the summer meeting near the Lee farm in Oxford County that year. So the Lee family decided to have the Sunday social dinner at Canada's Outdoor Farm Show site near Woodstock.

> There is still enough unmarked wood left, front to back, to accommodate the signatures of at least another seventy presidents— until 2087. That is quite a legacy!

Ginty provided a warm welcome to the guests, and then presented the paddle to President Hoftyzer, while reminding him of his responsibility to keep the boat heading in the right direction. From that year forward, every outgoing OSCIA president has signed the paddle and presented it to the incoming president at the annual meeting. Andy Graham and I have calculated that there is still enough unmarked wood left, front and back, to accommodate the signatures of at least another seventy presidents—until 2088. That is quite a legacy!

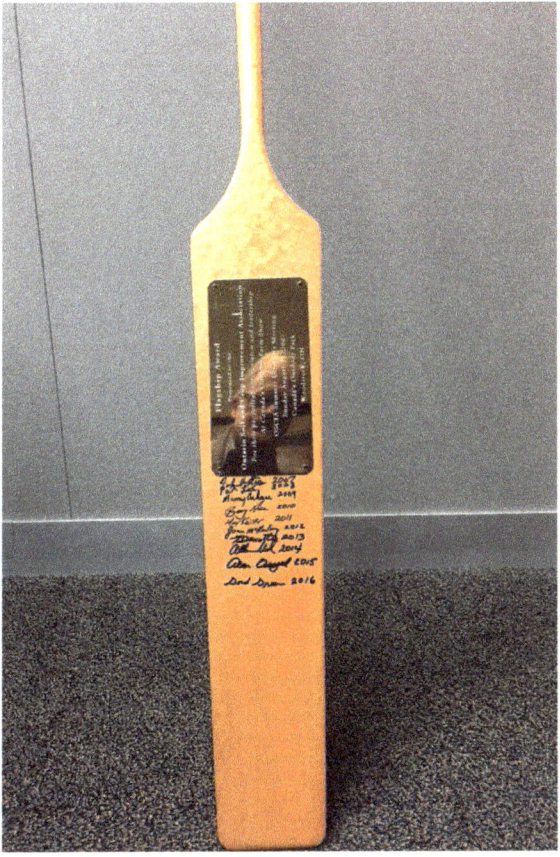

Fig. 11.1 **Flagship Award, profiled on the brass plate of the paddle with past-presidents' signatures below.** (Photo source: OSCIA)

Inclusiveness

Historically, the OSCIA membership reflected that, as in most agricultural organizations, men took a more active role in governance issues. However, this is changing quite dramatically. In fact, OSCIA's current strategic plan identifies gender imbalance as a weakness. The annual meeting in February 2018 included a panel discussion titled Success Stories of Women in Agriculture. I anticipate OSCIA's gender imbalance will be adjusted with strong leadership evident at all levels, beginning at the grassroots.

OSCIA's first female president was Joan McKinlay, from Grey County. Joan, her husband James and son Robert operate a cow-calf and cash crop farm at the top of Blue Mountain and Joan brought exceptional insight, experience, and leadership to OSCIA. Although it took OSCIA until 2012 to elect its first female president, OSCIA will undoubtedly continue to move toward gender parity on its board of directors in the coming years. Of the administrative staff in the Guelph office, over 70 per cent are female.

> 68 per cent of undergraduates enrolled in the University of Guelph's Ontario Agricultural College (OAC) are now female

Reflecting the demographic shift seen on university campuses, 68 per cent of undergraduates enrolled in the University of Guelph's Ontario Agricultural College (OAC) are now female.[151]

Furthermore, technology in agriculture is rapidly advancing to remove many of the physical demands and shifting the focus to skill, expertise and experience. According to the 2016 Canadian Census of Agriculture, 28.7 per cent of Canadian farm operators are women.[152] Strong leadership is coming from organizations such as the Ag Women's Network (AWN), whose vision is to provide "a dynamic agriculture industry which celebrates diversity and allows individuals to reach their full potential."[153]

According to Jen Christie, chair of AWN:

"Recognition of the value of diverse leadership, (inclusiveness) is leading to a greater representation of women on boards and in the corporate offices but there is more to be done. Groups like the Ag Women's Network are

---

151   Personal communication with OAC Dean, Dr. Rene Van Acker and Harold Rudy (January 19, 2019).

152   2016 Census of Agriculture, Younger operators and women make up a larger share of farmershttp://www.statcan.gc.ca/daily-quotidien/170510/dq170510a-eng.htm (Accessed February 14, 2018).

153   Ag Women's Network. About. www.agwomensnetwork.com (Accessed January 15, 2018).

providing women with a forum to build their network, find mentors and access resources to become stronger leaders."[154]

It all begins at the local grassroots level. As a provincial initiative in 2016, the OSCIA board established a new staff position, hiring Brittany Roka as a development adviser. One of her tasks is to provide leadership training to local and regional associations to enhance communication and membership recruitment. Brittany is working with the associations to enhance efficiencies and engagement levels amongst members.

OSCIA is also becoming more ethnically diverse. In 2010, Barry Hill of Brant County became OSCIA's first president from a First Nations background. When Barry and his wife Cheryle hosted the summer meeting in 2009, we were introduced to a number of colourful traditional dances and celebrations in their community. Aside from his multiple master's degrees and experience as an engineer for Ontario Hydro, Barry also enriched OSCIA with his expertise as a teacher, knowledge as a farmer, and talents as a musician.

What is the State of Your Constitution?

What are the traits of a successful organization? For OSCIA, it started long before my time. OSCIA is incorporated under the Agricultural and Horticultural Organizations Act, RSO 1990. It has a constitution, which provides direction on how it is structured and governed. There would have been a constitution when OSCIA was formed in 1939 and it has been tweaked from time to time over the years. Currently, the constitution identifies fifty-three local county/district associations within eleven regions. Each of the local and regional associations has its own constitution modelled after the provincial body.

---

154   Personal communication with Jen Christie and Harold Rudy (January 15, 2018).

The Cream Rises to the Top

> *"My presidency was the period where we combined the county/district organizations into regional groupings, not without some controversy. The Regions have evolved into progressive, informative, and respected organizations today."*
>
> -Keith Black, 2006 OSCIA President (Huron County)

It is not my intention to wade into a constitutional analysis or even review the changes to OSCIA's bylaws over the years. I may not have originally understood the importance of this legal document, but I now view OSCIA's constitution as rock-solid. I would, however, like to point out its unique features and why OSCIA has withstood the test of time for close to eighty years.

Each of the fifty-three local associations has a say in who is elected to represent them as regional director on the provincial board. With eleven regional directors serving on the provincial board, all of the agricultural landscape across Ontario is covered. As the saying goes, the cream rises to the top. It is an honour and a privilege for each of the eleven directors to serve in this capacity. They are highly respected within their communities and have a solid understanding of agricultural technology, science, and farming's relationship to their surrounding environment. Many of them have participated in on-farm research trials and opened their farms to others, even non-members, for field days to profile research results. They are among the elite from their community and tend to be modest in their achievements with few political ambitions. Their focus is on production agriculture with a conservation ethic, and they certainly have a strong commitment to their communities.

I have noticed also that the OSCIA directors are strong environmentalists. There are hundreds of examples where they have quietly invested in stewardship practices and showcase their farms to share their achievements within their community. Often, they are profiled in the provincial farm media. The conservation work of Henry Denotter, 2013 president, has been profiled in over six feature articles in agricultural publications and on the front cover of at least three magazines. An Essex County grain producer, Henry is also an avid photographer and studied photography in college, coincidently with Doug Wilson, president and CEO of the Cambridge Butterfly Conservatory.

In his term as president, Henry made sure the OSCIA staff, executive, and their partners were treated to a tropical tour of the butterfly habitat and a wonderful Christmas dinner at the Butterfly Conservatory.

Administrivia

Administration is one of those necessary evils, which if done correctly, no one notices. But when administration flounders, all hell can break loose! Fortunately, OSCIA has a strong record of success in administration, keeping costs in line, its programs able to pass the muster under the watchful eye of many auditors. In fact, nothing about OSCIA's administration is trivial. Kudos to the proficiency of staff at many levels for dotting i's and crossing t's under the watchful eye of OSCIA's CFO, Julie Henderson.

There has been a guideline that administration costs should be 12 to 15 per cent of an overall contract, but that depends. The $20-million Land Stewardship Program was administered in the six to eight per cent range. In contrast, expenditures on a small communication project may be mostly administrative. Generally, the greater the funding amount, the lower the percentage of administration costs.

Head office in many organizations is often viewed with suspicion or disdain by those in the field. "What on earth do all those people do?" would be typical expression from the field. Countless jokes, movies, and books have poked fun at the stereotypical head office mentality. Any such references to inefficiency, illogical rationale, or bureaucratic bumbling do not apply to OSCIA.

When dispersing public dollars, all the checks and balances must be in place. No one wants to be embarrassed by any perceived misappropriation. The reporting requirements, performance measures and outcomes must be carefully monitored and recorded. There are many rules.

Lawyers have been quite helpful to staff along the way. Of course, all our contracts and agreements have been drafted by lawyers — government lawyers. You guessed it: they are looking after government interests. So if you want to do business with the government, accept the agreement on their terms or don't do business. The best OSCIA can do is practice due diligence conduct regular audits and provide transparency to our government partners.

The greatest factor that has kept OSCIA at the forefront of administrative excellence has been the outstanding working relationship with our partners. What does that mean? In the best of circumstances, OSCIA has been engaged at the program design stage. We have been fully transparent through the implementation and delivery, providing timely reports and identifying some of the shortcomings. In many cases, local associations have played a key role in communication and marketing the program. In some cases, they have been too effective, as demand for cost-share funds far exceeded supply.

The employees at OSCIA have never been complacent. Andrew Graham, my close colleague in senior management and now executive director of OSCIA, establishes very high standards. The Ontario farm community has been fortunate for his tenacious attention to detail and ability to multi-task through reports and communications with clarity. With each program launch, there are high expectations that the objectives will be met, and the wind-down will occur on time and on budget. Interim and final reports, too, must happen like clockwork. I have been fortunate to work with exceptional colleagues at all levels from OSCIA's provincial office in Guelph to the field staff in all fifty-three counties/districts in rural Ontario, and colleagues in government and other agencies who have been generous with their time and contributed to our successes. As a senior manager, I could not have asked for more.

Also, in a governance model where staff reported to the board and where the board is ultimately responsible for the organization, I have been fortunate that the board allowed staff to manage day-to-day activities without interference. Each of my thirty-plus presidents, in their year at the helm, became my key contact. With some, I would be on the phone every day or two. Others were content to let me use my judgment to keep them informed on an as-needed basis. Of course, if an issue that had any political implications arose, they would receive a heads up.

OSCIA attempts to remain as non-political as possible. Polite visits with ministers of agriculture were always well-received. On occasion when major funding contracts were running out, or renewal was vaguely promised and cash flow was at stake along with staff livelihoods, our political pressure became more persistent. Transitions between generations of programs with multimillions of funds at stake could become stressful for staff and the board.

Thankfully, the basket of goodies usually arrived on time. Although a formal agreement may not yet have gone through all the legal and signing channels, there were always sufficient emails in place and historic good will to forge ahead. I would call this a calculated risk that we were willing to take. As OSCIA established deeper reserves, it was less nerve-wracking to invest in human resources and program development without having all the pieces of a new program in place first. OSCIA would sometimes keep a program afloat for six months or so before the first cheque found its way to our bank account. Cash management is so critical and CFO Julie Henderson is a master, claiming it as her specialty.

Don't Leave the Soil out of the CIA

What's in a name? On a few occasions, I've heard humorous comments about the "CIA" in OSCIA. Of course, the CIA reference was to the U.S. Central Intelligence Agency. Perhaps OSCIA could mean the Ontario Soil and Crop Intelligence Agency, since one of our major functions has been to inform, educate, and train farmers on best practices for managing soil and crops. The intelligence part has been assumed because OSCIA relies on the best and latest research to support its projects.

The name itself has been discussed on a few occasions over the years. It is a long name and doesn't exactly roll off the tongue. There is merit in a shorter handle such as 'Soil and Crop Ontario' or 'Soil and Crop' as we are commonly known in the farming community. In general, the acronym OSCIA is the most common among our colleagues; a few outsiders pronounce it as "OSKIA" with a hard rather than soft 'c'.

Actually, the word "soil" wasn't part of OSCIA's name until 1952. From its formation in 1939 until then, it was called the Ontario Crop Improvement Association. Under the leadership of 1952 president and Essex County farmer George Wallace, the name was changed to the Ontario Soil and Crop Improvement Association "in recognition of the inter-relationship between soils and crops."[155]

---

155   Dyszuk, B. "Two Blades of Grass Where There Was One Before." (1989) p. 3 (OSCIA Archives).

It will be left up to the next generation to ponder the name. In addition to the historic focus on helping farmers strive for higher productivity from their soils and crops, the OSCIA mandate now includes water and air, as well as biodiversity, pollinator health, and species at risk. What about our workshops on food safety and farm business planning? It all falls under the category of "sustainability" and that topic warrants a whole chapter later in this book. I currently participate on a provincial and national working committee to sort out how the farm, food and other ecological services can be tied together, often described in government circles as social license, or social contract. Perhaps OSCIA could change its name to the Ontario Social Contract Interpretation Agency—the acronym could stay the same!

Tossing the Sheaves

> "OSCIA working with partners to get the job done, in my mind, has always been paramount to OSCIA's success, and perhaps existence."
>
> -Jim Fisher, 1997 OSCIA President (Bruce County)

Figure 11.2 OSCIA Retired Logo Illustrating a Sheaf of Grain.
(Source: OSCIA)

Until 1989, the logo of OSCIA had been a sheaf of grain. Those of us who grew up with binders and threshing machines have lots of experience in stacking sheaves. A sheaf is a clump of grain and straw cut six inches from the ground and tied into a secure bundle with twine using a machine called a binder. Sheaves are then stacked onto their end, into stooks to dry before pitching them into a wagon for transport to the barn for threshing the straw from the grain. Stooks are still visible today in the fields around the Old Order Mennonite farms, but most OSCIA members no longer stook their grain. Sheaves were a legacy of the past, so a new logo focusing on the future was developed in preparation for OSCIA's golden anniversary in 1989.

> The outreaching hand is indicative of a soil caretaker, seeking new and innovative cropping methods

The new design featured an outreached hand holding a plant against a background representing a piece of machinery working a field in a grid pattern. The outreaching hand is indicative of a soil caretaker, dedicated to innovative cropping methods. In modern agriculture, grid co-ordinates are used to map and test soils, and precisely manage planting, spraying and harvesting. All this thanks to the same GPS technology found in our cars and smartphones. The symbolic hand is seeking new ideas to be tested through the collaboration of members and extension staff. The new logo remains and is as relevant today as when introduced in 1989.

Figure 11.3 – OSCIA's new logo in 1989.
It illustrates the outreaching hand and grid pattern of the field. The slogan "Grassroots Innovation Since 1939" was added circa 2011 (Source: OSCIA)

Regional Communication

> *"One of my major accomplishments was to help the OSCIA align the board members with the regions and the regional communication co-ordinators to better serve the OSCIA members."*
>
> -Frank Hoftyzer, 2007 OSCIA President
> (Peterborough County)

In 1999, OMAFRA announced that it was closing a number of its county/district offices and consolidating under a regional structure. The rationale given was that bricks and mortar were less important than the expertise of staff and the level of services. Websites and electronic communication were expanding at lightning speed so information could be accessed and communicated from any location. Regional offices encompassing multiple counties/districts were quickly becoming the norm.

This was a significant announcement for OSCIA as the secretarial services for the county/district OSCIA offices were typically carried out by a government employee. Further, the government office organized, wrote, printed, and mailed a newsletter several times per year to inform the farm community about new scientific developments, upcoming meetings, and happenings in and around their local area. Government staff also co-ordinated meetings and helped set up the agendas, guest speakers, locations, and catering. With the new restructuring at OMAFRA, these services would no longer be provided.

There certainly was some drama at the time. Change is always difficult but to paraphrase Plato and few other notables, the measure of humankind is how change and challenges are tackled. The OSCIA executive committee and board of the day, led by 1999 president Allan Yungblut, took on the challenge.

> *"I would say that forming the regional associations was the most major accomplishment. I have enjoyed watching the association grow stronger with activities like creating our own very professional newsletter, regional conferences (e.g. FarmSmart), and stronger projects."*
>
> *-Allan Yungblut, 1999 OSCIA President (Niagara North Region)*

For starters, OSCIA received funds to set up the regional infrastructure to jibe with OMAFRA's reorganization into eleven regions. Crop technology manager Brent Kennedy helped reorganize the OMAFRA field staff to match the OSCIA regions. In addition, the OMAFRA field staff would take on provincial lead positions with specialties in areas such as corn, soybeans, wheat, colored beans, forages, and soils.

Modus Operandi

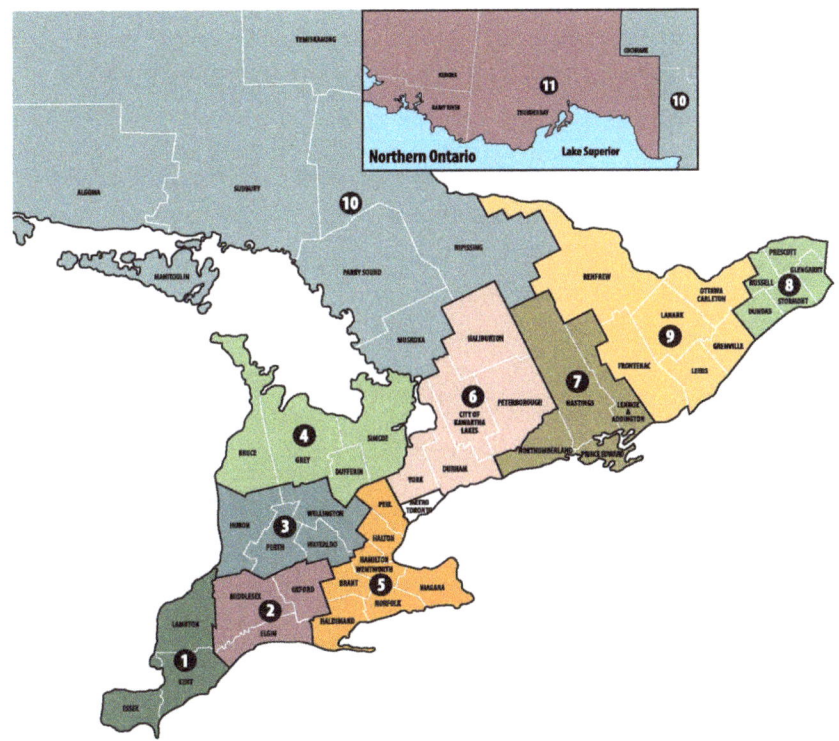

Figure 11.4 – OSCIA Regional Associations. (Source: OSCIA)

1 St. Clair	4 Georgian Central	7 Quinte	10 North Eastern Ontario
2 Thames Valley	5 Golden Horseshoe	8 Eastern Valley	11 North Western Ontario
3 Heartland	6 East Central	9 Ottawa Rideau	

The eleven new OSCIA regions were a great match for OMAFRA but there was some consternation over where to draw the regional boundaries. Historically, some clusters of counties and districts had already been collaborating on projects. This was the case in northern Ontario, where Rainy River, Thunder Bay, and Kenora formed the North-West Region. A similar Region for Temiskaming, Cochrane, Nipissing, Parry Sound, Muskoka, Sudbury, Manitoulin, and Algoma formed the North-Eastern Region. One of the challenges in the North is distance: it's a seven-hundred-kilometre trip one-way for anyone needing to travel across the region from Sault St. Marie to Cochrane. That's almost as far as from Guelph to Washington D.C.! Of course, Guelph to Rainy River is 1,856 km travelling through Ontario, as far as from Guelph to Jacksonville, Fla. Ontario is huge geographically!

Figuring out the boundaries in southern Ontario was more challenging. There were some informal working arrangements, but the provincial board left it up to the local county associations to figure out their partners. The provincial position was clear. There would be equal funds given to each region for them to acquire a part-time regional communication co-ordinator, whose primary responsibilities would be to publish four newsletters per year, replacing the newsletters published by the OMAFRA offices that were closing. In the end, the regions organized their own structures, which included from three to eight counties each. There were no orphans.

An unusual twist occurred in Grey County, where the local farm organizations requested their own agricultural office after the closure of the government office in Markdale. Under the leadership and vision of Ray Robertson, one of OSCIA's field staff at the time, Grey Agricultural Services was opened on May 1, 2000, to serve the farm and agri-business community in the region. The local farm organizations were extremely supportive to set up an arrangement with the county of Grey to provide an administrative office and board room space. That office is the headquarters for the Georgian Regional Soil and Crop Improvement Association and administers Grey-Bruce Farmers' Week, Ontario Biomass Co-op, Ontario Forage Council, Ontario Hay Listings, and the Grey-Bruce Alternative Land Use Services (ALUS) program.[156]

Forming the OSCIA regions in partnership with the OMAFRA crop technology staff was a win for both. The OMAFRA crop leads became focused experts in their specific crop or soil specialty. They gained prominence across the province and were in high demand for setting up on-farm research trials, analyzing and publishing results, plus attending numerous meetings and field days across the province.

For OSCIA, the regional communication co-ordinators have taken on significant prominence also. In recent years, under the leadership of

---

156 Grey Agricultural Services, http://www.greyagservices.ca/ (Accessed January 3, 2018).

Cathy Dibble, the newsletters became professional in appearance. Design, production, printing, and distribution became more efficient to save costs. Advertising by local agri-business establishments helps to defray costs and enhance the newsletters. A three-way partnership involving the local region, the OSCIA provincial office and OMAFRA *Crop Talk* provides the content for an informative package. The contributions of our eleven regional communication co-ordinators over the past sixteen years has been outstanding.

We now have exceptional regional conferences throughout the winter every year: the South-West Agricultural Conference in Ridgetown (SWAC); FarmSmart in Guelph; the Eastern Ontario Crop Conference; Grey-Bruce Farmers Week; East-Central Farm Show; the North Eastern Trade Show and Conference; Thunder Bay Spring Conference; and many smaller or local workshops.

Performance Anxiety

The first graduate course I took in 1989 was in program evaluation. That experience was invaluable as I was thrust into a totally new program delivery model, not without some criticism from onlookers. Therefore, I felt it was my duty to take a critical look at the important indicators of success, or flag deficiencies that could be addressed in any future delivery. On the political side, an easy measure was the number of letters or phone calls complaining about a program or expressing displeasure to a local MPP or the agriculture minister. When it came to local committees of farmers leading delivery of the LSP in their counties/districts, there were few complaints.

Documenting performance measures is one of the greatest challenges for administrators in any organization or business. It's perhaps even more challenging when spending taxpayers' dollars, as the scrutiny can be much more rigorous.

Right from the get-go for LSP, we set up our administrative structure in preparation for audits. OSCIA retained an accounting firm that conducts annual financial audits, so for us, intense scrutiny is part of the routine. But there is more.

Hearing that your office or program will be audited by government can lead to anxiety. An audit or evaluation can compare program objectives with what was actually achieved. There are always questions of interpretation. Program guidelines generally clarify eligibility and non-eligibility. What about the outliers? For a series of federal programs under the Agricultural Policy Framework, AAFC staff in particular established standardized documents called Record of Decision forms that were extremely helpful months or even years down the road when an auditor was looking for justification for decisions.

> Conducting an audit every six months would meet the requirements of both government contracts and OSCIA accountability to members

I also kept a folder, which I called an audit file, to record details of less formal decision-making. If I came across a unique or unusual circumstance that required a judgment call, I wanted to back up my decision if, at some point in the future, I was required to justify my rationale. In earlier years, it was a paper file. Since about 2003, I've kept an electronic version where I have stored emails or other electronic documents for future reference.

OSCIA now conducts financial audits every six months. In recent years, OSCIA typically has twelve to fifteen government contracts on the go, most with termination dates of March 31, after which, logically, an audit is conducted for many agreements.

However, March 31 is not a logical date for farm organizations to have a year end. Typically, annual meetings are required within two to four months after year end. This would peg OSCIA's annual meeting sometime over the summer. It would be impossible for OSCIA delegates to attend a summer annual meeting when their attention is drawn to their cropping duties back at the farm. These meetings must occur in the winter when farmers have time to attend.

Historically, the OSCIA fiscal year end had been Nov. 30, as this was typically the end of field activities, when bills could be paid and an audit conducted prior to the annual meeting in February. But the November date became impractical as the number of program contracts increased and the terms grew more complex, with more lead time required for bank reconciliations, audit, and review.

Modus Operandi

With the government year end of March 31 and OSCIA's year end at November 30, we were not in sync. Many options were discussed by the OSCIA board leading up to the spring of 2006, including all the disadvantages of moving OSCIA's year end to match government's March 31. Our compromise? We would move OSCIA's year end to Sept. 30, thereby creating six-month intervals between audits. Conducting an audit every six months would meet the requirements of both government contracts and OSCIA accountability to members. This also reduced performance anxiety to a considerable degree. It has been working marvellously ever since.

> Farmers have a unique ability to share with a high degree of credibility, their goals, ethics, and commitment to leaving their land in better condition than it was when they started farming

During my career, there were at least four external audits conducted by government or third-party auditors hired by government. These audits delved much deeper than finances. Extensive teaching opportunities occurred when a junior consultant from downtown Toronto dove into the technical details of conservation tillage equipment, no-till drills, manure spreaders, or GPS systems. There were often puzzled looks as they quizzed us about specific terms such as the definition of a no-till drill.

One of my best experiences with auditors was during the Land Stewardship Program, when the auditors visited the farms to see the conservation equipment in the farm sheds and trudged into the fields to observe the purpose of erosion control structures. They spoke directly to producers to learn the impact of our cost-share program on their farms. When asked, farmers have a unique ability to share with a high degree of credibility their goals, ethics, and commitment to leaving their land in better condition than it was when they started farming. A multitude of these stories and proud accomplishments improve understanding beyond the farm gate. Farmers are always improving their soil management, always fixing.

The farm visits satisfied the LSP auditors in 1989. In fact, that was the first real test of OSCIA's legitimacy as a program delivery agent. Funds were well targeted, they made a difference, clients were happy, and the public received great value for their investment in conservation improvement. We in the farm community need to share more of our good news stories!

Evaluating OSCIA's communication efforts is also challenging. OSCIA is an organization based on communication activities. What are the indicators of success for communication? The obvious is to record the number of newsletters, field days, workshops, and bus tours. Social media provides a whole new set of metrics to track our efforts aimed at engaging members and stakeholders.

> Some performance measures are now built into our merit-based eligibility requirements

The boards of directors in each county/district association intuitively evaluate their successes (and failures) from one year to the next. Improvement in crop yield or reduction in costs are front and centre to local membership. This is essential for financial survival, but the almighty dollar doesn't drive all decisions. There is a strong stewardship ethic among the rural landowners. Social media is providing the means for low-cost communication. The vibrant and dynamic stories are changing the face of agriculture.

Performance Anxiety over Environmental Improvement Measures

We still struggle with obtaining hard numbers about how our programs impact the environment. How many tonnes of soil did we prevent from eroding into adjacent streams? What was the reduction in phosphorus in Lake Erie because of the LSP? How much silt is collected in a vegetative filter strip along a stream after we provided cost-share to a landowner for planting the buffer? How many manure spills were prevented because of improved training on best practices for storage, transportation, and application? How many endangered species did we prevent from dying, or did we in fact increase their population? These are difficult measures and OSCIA and partners continue to struggle with obtaining meaningful measures. My colleagues have often heard me rant, "How does one measure something that didn't happen?" The rationale is that it is much more cost-effective to prevent something from happening than to clean up the mess after it happens. We are getting better at documenting performance measures but it does come with higher administrative costs. In the early days of

EFP development, environmental performance measures were discussed. Policymakers however, wanted to see maximum dollars transferred into the farm community to satisfy the farmers' action plans, rather than spend program funds on expensive studies and reports. Some performance measures are now built into our merit-based eligibility requirements.

Charitable Status

Most environmental organizations are registered as charities to attract donations whereby a receipt can be provided for income tax deductions. On a number of occasions, we considered the benefits of establishing charitable status for OSCIA. For some funding agreements, only an organization with charitable status would qualify; in such cases, OSCIA had the funds channeled through a charity such as the Lake Simcoe Region Conservation Authority (LSRCA). LSRCA then executed a sub-agreement between LSRCA and OSCIA.

What about donations to OSCIA? Providing charitable receipts to donors could be advantageous to OSCIA. Our board members suggested that with the current level of capital accumulation in farm land, there may be interest by farmers in directing some of their estate to a reputable charitable organization that has programs focused on conservation and soil improvement.

We began investigating options in 2009. In order to qualify as a charity in the eyes of the Canada Revenue Agency (CRA), an organization must meet one or more of the following purposes:

- the relief of poverty
- the advancement of education
- the advancement of religion, or
- other purposes beneficial to the community in a way the law regards as charitable.[157]

---

157   Government of Canada, Canada Revenue Agency website, "What are the general requirements for attaining and maintaining charitable registration?" https://www.canada.ca/en/revenue-agency/services/charities-giving/charities/policies-guidance/policy-statement-029-research-a-charitable-activity.html?wbdisable=true#toc2 (Accessed February 26, 2018).

A formal application would have to be submitted to CRA. In discussion with a CRA representative, OSCIA would also have to seek approval from OMAFRA, since OSCIA is incorporated under the Agricultural and Horticultural Organizations Act, R.S.O., 1990, which is administered by OMAFRA. A separate board would be required along with separate financial statements with annual audits. We consulted with other agricultural organizations that have already established charitable status and we learned that it would cost about $30,000 annually to operate a charity, separate from the parent organization. In the end, an alternative approach was pursued to achieve most of OSCIA's objectives.

> Through the generous contributions of OSCIA Founding Partners, the scholarship was launched in 2015 and will be available until at least 2019

Rather than submit an application to CRA, the OSCIA board agreed to align with another charity with common interests to OSCIA. Dr. Rob Gordon, then-dean of the Ontario Agricultural College at the University of Guelph, opened the door for this option. OSCIA established a Soil and Crop Sustainability Fund, which includes a graduate scholarship to support expanded research on soil health issues. Tax receipts would be provided by the university to donors. Where tax receipts were not required, OSCIA organized to administer an internal fund to be used for applied research at the board's discretion.

Soil Health Graduate Scholarship

> *"We're proud to have had the opportunity to initiate a scholarship at University of Guelph for students in agronomy, supporting the principles and objectives of OSCIA. The tradition has been carried forward by OSCIA at the graduate level. Cheryle and I continue with an undergraduate level scholarship in agricultural studies."*
>
> -William Barry Hill, 2010 OSCIA President (Brant County)

The Soil Health Graduate Scholarship provides $10,000 per year to graduate students focused on soil health or soil quality research. Recipient selection is completed internally by the University of Guelph. OSCIA will be involved in collaboration and communication of research where possible. Through the generous contributions of OSCIA Founding Partners, the scholarship was launched in 2015 and will be available until at least 2019.

OSCIA Internal Fund

The OSCIA Internal Fund will support on-farm applied research for soil health, as well as outreach and communication of results. Local or regional soil and crop improvement associations may be able to apply for funding to support applied on-farm research. Funding is anticipated to be available starting in 2017.

Figure 11.5 Founding Partners of the OSCIA-Sponsored Sustainability Fund (Present at OSCIA AGM, February, 2015), Standing (L to R): Henry and Cathy Denotter, Richard Sovereign, Steven and Sandra Eastep, Cathy Dibble, Ruth and Don Hill, Laura Green, Barry Hill, Betty and Allan Brown, Gord Green, Ann and John Benham, James McKinlay, Greg Kitching, Robert McKinlay; Seated (L to R): Carl and Valerie Bolton, Harold Rudy, Joan McKinlay. (Photo source: OSCIA)

Jaclyn Clark became the first recipient of the Soil Health Graduate Scholarship in 2015. Jaclyn investigated "Cover crop planting into standing corn." Her research involves inter-seeding red clover and ryegrass into standing corn in early summer. By the time the corn is harvested, the cover crops will have had a head start on growth, protecting the soil and capturing leftover nutrients. The goal is to improve soil health and water-holding capacity for enhanced drought resilience.

Figure 11.6 Cover Crop Planted into Standing Corn.
Research was conducted as part of the Soil Health Graduate Scholarship (Photo source: University of Guelph)

The 2016 Soil Health Graduate Scholarship was awarded to Jordan Graham, whose research is focused on the "Influence of herbaceous biomass crops on soil organic carbon levels in soils across Ontario."

The 2017 recipient Pedro Ferrari Machado will be working as a PhD candidate with scientist Dr. Claudia Wagner-Riddell. His research will evaluate the long-term effects of tillage and crop rotation practices on emissions of nitrous oxide, a potent greenhouse gas.

The OSCIA board is also open to working with other universities or charities to receive donations in support of work that meets the common goals of soil and crop management.

The OSCIA Soil Champion Award

> Steel-toed shoes have not yet been a requirement at OSCIA's annual meeting

Over the years, many soil champions have continued to demonstrate exemplary commitment to soil conservation. In 2014, long-time conservation advocates Lillie Ann Morris and Don Lobb approached OSCIA about sponsoring an annual Soil Champion Award. OSCIA jumped at the opportunity and quickly developed criteria, nomination forms, and a promotion plan. The new Soil Champion would be announced each year in February at OSCIA's annual meeting.

To be eligible for the Soil Champion Award, an individual must be a resident of Ontario or have contributed to soil management in a way that directly influences improved soil health and sustainable crop production in the province. Sustainable soil management practices may be defined as those that:

- Make the most efficient use of nutrients;

- Support systems with no net loss of organic matter and soil aggregate stability;

- Build the population and diversity of soil organisms; and

- Effectively manages surface water to support reduced tillage systems.[158]

We knew there would be lots of candidates for the honour, but what would we provide as an award? A sports trophy? Not likely! Maybe a chrome-plated coulter (a disc-shaped farm implement that is used for seed placement on a no-till planter)? Apparently, these coulter-trophies are provided in some jurisdictions but the danger of dropping a twenty-pound, razor-sharp trophy

---

158 OSCIA Soil Champion. www.ontariosoilcrop.org/association/association-soil-champion-award/ (Accessed July 19, 2017).

on someone's toes negated that possibility. Steel-toed shoes have not yet been a requirement at OSCIA's annual meeting.

We elected to proceed with a trophy made from a baseball bat with graphics and an engraved plate profiling each year's winner. Baseball and champions go together. I had seen trophy bats at a baseball bat factory in Cambridge, Ont. The KR3 Bats Company is a hidden gem. Locally owned by the Huehn family, KR3 Bats ships its hand-crafted products all over the world. Yes, some bats even go to the majors![159]

OSCIA Soil Champions

<u>2018 Soil Champion Dan Breen</u>

- Dan Breen is a third generation Middlesex County dairy farmer near Putnam, Ont., who began no-till in 1990 and hasn't looked back. His diverse crop rotation of over eight hundred acres includes corn, soybeans, wheat, alfalfa, and cover crops;

- Manure is applied only when the soil conditions are favourable;

- "A true no-till system is more than just not tilling, it is biodiversity, water retention, and nutrient recycling. As a farmer, I've had an opportunity to be a caretaker of this land, but I only have tenure for a blip in history. I hope I leave it in better shape than when I found it, and I hope my daughter and son-in-law will do the same."

Figure 11.7 Dan Breen, 2018 OSCIA Soil Champion
(Photo source: OSCIA)

---

[159] KR3 Bats. www.kr3bats.com (Accessed Aug. 28, 2017).

2017 Soil Champion Eric Kaiser

- Kaiser Lake Farms is an egg and field crop operation on the shores of both Hay Bay and the Bay of Quinte in Lennox and Addington County;

- Long-time advocate of bundling no-till farming, crop rotation, structural erosion control, systematic tile drainage, and diverse cover cropping;

- Frequent speaker and participant in events across North America. "Sustainability has many components, but the preservation of topsoil must be the final result."

Figure 11.8 Eric Kaiser, 2017 OSCIA Soil Champion with son Max
(Photo source: OSCIA)

2016 Soil Champion Tyler Vollmershausen

- Field crop producers near Innerkip in Oxford County;

- Soil management practices include strip till, no-till, and cover crops;

- Tyler is a strong advocate of soil health and regularly shares knowledge and experiences with others in farming;

- "Our focus is soil health. Keep the soil covered at all times and keep living plant roots in the ground. That's what is driving the biology below the soil line."

Figure 11.9 Tyler Vollmershausen, Vollmershausen Farms, 2016 OSCIA Soil Champion
(Photo source: OSCIA)

2015 Soil Champion Dean Glenney

- Field crop producer from Dunnville area in Haldimand County;

- Innovator of the fence row farming concept;

- Achieved astounding crop yields by building healthy soil structure on clay soils and attending to incredible microbial activity including bacteria and fungi;

- "You have to get the fundamentals right."

Figure 11.10 Dean Glenney (L.), 2015 OSCIA Soil Champion with 2014 OSCIA President, Allan Mol
(Photo source: OSCIA)

2014 Soil Champion Adam Hayes

- Soil Management Specialist, Ontario Ministry of Agriculture, Food, and Rural Affairs;

- Outstanding career accomplishments in promoting soil health across the agriculture industry;

- Leading expert in knowledge translation and transfer;

- Played pivotal role in the success of the annual Southwest Crop Diagnostic Days and South West Agriculture Conference held at Ridgetown Campus, University of Guelph.

Figure 11.11 Adam Hayes, 2014 OSCIA Soil Champion
(Photo source: OSCIA)

It's Official – Ontario Now Has a Soil Designation!

Ontario has an official flower, the white trillium; an official bird, the common loon; and an official tree, the white pine. What about an official soil? Manitoba, Nova Scotia, New Brunswick, Prince Edward Island, Quebec, Alberta, and British Columbia have their official soil designations. All U.S. states have an official soil designation,[160] so why not Ontario?

> The Guelph Soil Series is a good choice because it is one of the most productive soils in Ontario

It DID happen. On November 30, 2015, Jeff Leal, Ontario's Minister of Agriculture, Food, and Rural Affairs, declared the Guelph Soil Series as Ontario's official soil with a celebration that included the Hannam family at their Woodrill Farms near Guelph, Ontario.[161]

Figure 11.12 Ontario's Official Soil Designation Ceremony – Guelph Soil Series
(L to R): Sandra Hannam, Hon. Jeff Leal Minister, OMAFRA, Alan Kruszel, 2015 OSCIA President, with Greg Hannam and Peter Hannam. (Photo source: OSCIA)

---

160 USDA, Natural Resources Conservation Service, "Soils." https://www.nrcs.usda.gov/wps/portal/nrcs/detail/soils/edu/?cid=stelprdb1236841

161 OMAFRA, News, "Ontario Designates Guelph Soil Series Official Provincial Soil," https://news.ontario.ca/omafra/en/2015/11/ontario-designates-guelph-soil-as-official-provincial-soil.html (Accessed December 28, 2017).

I reckon it took until 2015 because Ontario has close to three hundred soil series so there would be lots of debate on which one to select. The Guelph Soil Series is a good choice because it is one of the most productive soils in Ontario. Additionally, back in 1914, the Guelph Soil Series was one of the first group of soils in Ontario to be described by the Ontario Soil Survey.[162]

The Guelph Soil Series encompasses more than 70,000 hectares of productive loams, sandy loams, and silt loams throughout the counties of Brant, Dufferin, Oxford, Perth, Lambton, and Wellington, as well as within the regions of Halton, Waterloo, and the city of Hamilton. OSCIA 2015 President Alan Kruszel participated in the official ceremony with the Hannam family, owners of Woodrill Farms. The engraved ceremonial shovel from the event is mounted in OSCIA's office alongside OSCIA's official paddle and a sample of the Soil Champion trophy baseball bat.

Figure 11.13 – OSCIA's Trophy Wall - The ceremonial shovel used for the designation of the Guelph Soil Series as Ontario's Official Soil. The Soil Champion trophy baseball bat waits for the next Soil Champion award ceremony. The Flagship Award paddle is signed by outgoing OSCIA presidents and presented to the incoming president, advising him or her to paddle the boat in the right direction. (Photo source: OSCIA)

---

162   Ibid.

Having a strong administrative unit with competent staff has been one of the keys to attracting new opportunities for program collaboration, expanding OSCIA's activities, and providing support to the local network of county/district/regional affiliates across Ontario. Bringing together applied science to focus on research results continues to fuel OSCIA's mission to "facilitate responsible economic management of soil, water, air, and crops through development and communication of innovative farming practices." Although this text provides a provincial perspective on soil fixing, another large book could be written about on-the-farm testing and innovation where grassroots members and their neighbours seek continuous improvement to soil management and water protection.

Chapter 12

ON-FARM RESEARCH AND COLLABORATION

As we strive to fix soil, it is apparent that agricultural practices are all interconnected. Some might suggest that farms are one big organism, so that when you poke it on one side, it extends outward on the opposite side. Nothing can be done in isolation. Extensive scientific research is the only way to tie our best practices together to guide and direct. OSCIA's special niche has been to conduct on-farm research where farmers learn first-hand how to adopt new developments.

> Everything we eat has been genetically modified; it's just a question of which plant breeding method was used

Is There an Elephant in the Room?

There may actually be a herd of elephants in the room. I'd like to introduce you to a few that I have met. The first one that few want to acknowledge is how the term genetically modified organism or GMO is misused, misrepresented, and misunderstood. It is important to clear the air on terminology as it relates to soil care and protection, specifically to weed control and its relationship to genetically engineered seeds that facilitate the adoption of no-till planting.

According to the Canadian Food Inspection Agency (CFIA), "an organism, such as a plant, animal, or bacterium, is considered genetically modified if its genetic material has been altered through any method, including conventional breeding."[163] There are many forms of genetic modification: genetic engineering (GE), transgenic, living modified organism (LMO), mutagenesis, plant with novel traits (PNT), selective breeding, and more. This chapter is not intended to provide a complex lesson in genetics (and believe me, it is complicated), but to recognize that when conducting research in agriculture, it is misleading to talk of two camps, GMO and non-GMO. Why? Because everything we eat has been genetically modified, it's just a question of which plant breeding method was used.

> The development of engineered seed to simplify weed management and save soil was one of the greatest problem-solving breakthroughs for farmers in the past thirty years

The sweet potato, for example, is a crop in which a transgenic gene transfer involving *Agrobacterium* occurred naturally thousands of years ago through one of the same processes used in modern plant breeding today.[164]

Is it okay to consume food originating from transgenic gene transfer if it occurred randomly by accident thousands of years ago, but not food modified by a plant breeder who used the same technique with precision, extensive screening, testing, and scrutiny by a regulatory process to ensure it is safe for humans and the environment?

This debate introduces another elephant in the room: the argument that modern transgenic development is driven by greedy, profit-motivated multinationals. I have a friend who often waxes eloquently on the profit motivation of our largest seed companies until I remind him that we contribute to

---

163 Modern Biotechnology: A Brief Overview. Canadian Food Inspection Agency. www.inspection.gc.ca/plants/plants-with-novel-traits/general-public/overview/eng/1337827503752/1337827590597 (Accessed June 8, 2017).

164 Kyndt, T., D. Quispe et al. "The genome of the cultivated sweet potato contains *Agrobacterium* T-DNAs: An example of naturally transgenic food crop." Proceedings of the National Academy of Sciences of the United States of America (PNAS). (March 16, 2015). http://www.pnas.org/content/112/18/5844.full.pdf (Accessed June 19, 2017).

profit-driven motives if we own a cellphone, watch TV, purchase automobiles, or just about anything else. Additionally, I remind him that his choice of car brand labels him as a supporter of the most profitable auto manufacturer in the world. Perhaps he should purchase his next car from a manufacturer that is less profitable.

> He described how conventional tillage contributes to deteriorating soil conditions, less resilience to climate change and the problem of carbon dioxide emissions

So when it comes to labelling, it is downright dishonest to paint food as simply GMO or non-GMO, black or white. How does this relate to soil fixing and advances in soil conservation? The development of engineered seed to simplify weed management and save soil was one of the greatest problem-solving breakthroughs for farmers in the past thirty years. Profit-driven or not, it has been embraced by many farmers as a tool that allows them to be better soil managers with the adoption of no-till.

But not all farmers have embraced the use of genetic engineering technology to simplify weed control in no-till for soil protection. A few farmers are researching alternatives including roller-crimping techniques that also leave the soil with a protective cover. This involves planting suitable cover crops such as rye, then flattening the mature grasses with a roller-crimper machine and planting the crop into the flattened mulch. The added biomass provides cover for seedlings while enriching the soil and providing natural weed suppression that may reduce and even eliminate the need for herbicides. More research is required for these methods to be proven reliable year after year. What will the future hold and how common will this become? Stay tuned to The Soil Fixers sequel to find out.

When farmers began to adopt no-till planting to prevent soil erosion, one of the greatest challenges was weed control. Traditional forms of weed control included plowing, cultivation, disking, and mixing of the soil to disrupt the weed cycle. However, as discussed previously, excessive plowing and tillage may leave the ground bare and exposed to the elements, which greatly increases the risk of soil erosion. Additionally, overworking the soil destroys pores and microorganisms, adding to carbon loss. This point was driven home clearly by Dr. Don C.

Reicosky, a retired soil scientist from the USDA Agricultural Research Service, when he spoke at the Summit for Canadian Soil Health at Guelph in August 2017. He described how conventional tillage contributes to deteriorating soil conditions, less resilience to climate change and the problem of carbon dioxide emissions. Here's his chart to illustrate:

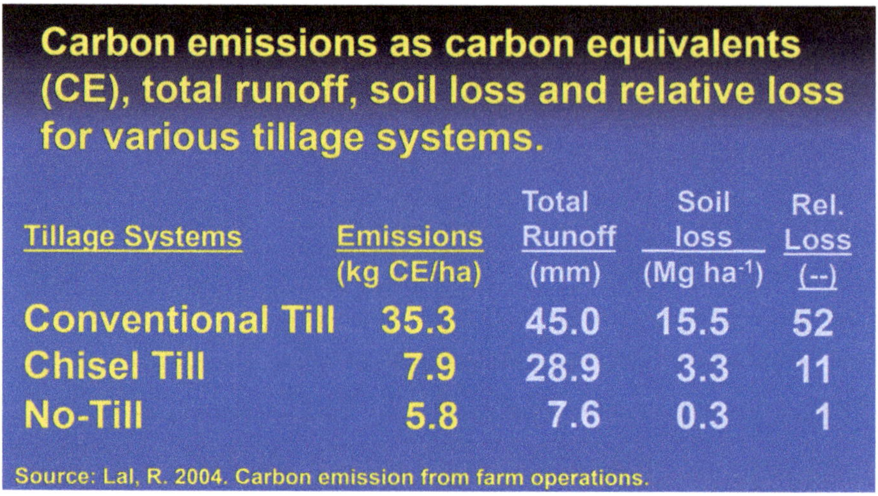

Carbon emissions as carbon equivalents (CE), total runoff, soil loss and relative loss for various tillage systems.

Tillage Systems	Emissions (kg CE/ha)	Total Runoff (mm)	Soil loss (Mg ha^{-1})	Rel. Loss (--)
Conventional Till	35.3	45.0	15.5	52
Chisel Till	7.9	28.9	3.3	11
No-Till	5.8	7.6	0.3	1

Source: Lal, R. 2004. Carbon emission from farm operations.

Figure 12.1 Conventional tillage exacerbates problems of soil and water runoff and produces nearly seven times more carbon emissions compared to no-till management.[165]

No-till planting relies on herbicides for weed control. Some traditional herbicides required mixing with the soil to be effective—impossible under no-till. Others caused plant stress, which could reduce yields. Farmers were desperate for new options. Along came a herbicide that would knock out a broad range of weeds and grasses, result in little stress to the plants, give season-long control and was safe to use with little impact on the surrounding environment. This tool was badly needed to complete the package for conservation cropping. Glyphosate, first developed and introduced into the market

---

165   Reicosky, D. C. "Carbon: The Synergy Element in No-Tillage and Cover Crops." www.soilcc.ca/news_releases/2014/congress/019%20Carbon-The%20Synergy%20Element_%20Don%20Reicosky.pdf Emissions source: Lal, R. 2004, (Accessed Aug. 29, 2017).

as Roundup in the 1970s, was such an herbicide. It became one of the most effective options for effectively cleaning fields of growing weeds to give the planted seeds a clean start under no-till without working the soil. Farmers, researchers, and extension staff were familiar with its benefits. Frankly, the traditional registered herbicides did not work well for no-till farmers so when glyphosate resistant soybeans were introduced in the mid-1990s, it was a godsend. It cleaned up the crops, was low cost, and safer for humans and the environment than previous products.

> This course and the EFP program promote best practices such as crop and herbicide rotation to prevent or minimize the risk of weed resistance

When Dekalb approached OSCIA to collaborate with on-farm corn trials to test a new Roundup Ready corn variety, farmers knew of the benefits so the OSCIA board was keen to get involved. The Roundup Ready Corn Showcase ran for two years in 1998 and 1999. Quite simply it was an opportunity for farmers to learn how to test a corn variety that was resistant to glyphosate, a herbicide that would economically control most weeds and was less toxic than table salt.[166]

Weed resistance may be of concern but can be managed. Farmers can learn about safety, proper handling, and application of pesticides through the Ontario Grower Pesticide Safety Course.[167] This course and the EFP program promote best practices such as crop and herbicide rotation to prevent or minimize the risk of weed resistance.

The real ethical question for the OSCIA board in 1998 was whether they should align with any private company to conduct on-farm research trials. But since its inception in 1939, the organization has encouraged farmers to test new seeds and cropping methods. OSCIA records do not provide

---

166   Fishel, F. et al. "Herbicides: How Toxic Are They?" Institute of Food and Agricultural Sciences (University of Florida, rev. February 2013) http://www.edis.ifas.ufl.edu/pdffiles/PI/PI17000.pdf (Accessed May 12, 2017).

167   Ontario Pesticide Education Program. www.opep.ca/certification/grower-pesticide-safety-course/ (Accessed Dec. 7, 2017).

extensive details about chemical use, but even ancient farmers exercised a myriad of products, such as sulphur, used 4,500 years ago by the Sumerians for insects and mites. We won't discuss mercury and arsenic, but some of those ancient products were brutal.[168]

> OSCIA would consider on-farm research for each technology based on its own merits

The OSCIA board suggested that if it was good for farmers, better for the environment, safer than previous products, supported by independent research, and legal for use, OSCIA would consider on-farm research for each technology based on its own merits. Any seed company or crop input supplier could approach OSCIA for an independent assessment of their product. That policy has served OSCIA well.

A close relative of that elephant in the room is whether OSCIA should accept ANY sponsorships from agri-business for field trials, meetings, or field days. A few might suggest, "Absolutely not! If we do, would we become shills, co-conspirators, pawns, and hostages of the greedy corporate world? Would industry support taint the results?" But neither farmers nor OSCIA are 'bought' by industry. I remind potential sponsors that this is a time to give back to the farm community as a gesture of goodwill for the revenues and profits they receive from farmer purchases throughout the year.

Since 2003, working as executive director, I've gone through an annual ritual each December to line up sponsors to defray the costs for the OSCIA annual meeting in February. Delegates attend from the fifty-three counties/districts across Ontario. OSCIA pays their travel and accommodation. Meeting at a respectable hotel in a central location like London, Toronto, or Niagara Falls is an expensive undertaking that can cost close to $100,000. Delegates enjoy guest speakers, reports about on-farm research, great networking among peers, and fine hospitality. I was pleased that for OSCIA's

---

168 Unsworth, J. "History of Pesticide Use." International Union of Pure and Applied Chemistry. (May 10, 2010) www.agrochemicals.iupac.org/index.php?option=com_sobi2&sobi2Task=sobi2Details&catid=3&sobi2Id=31 (Accessed Aug. 29, 2017).

seventy-fifth anniversary in 2014, we were able to acquire over $30,000 in sponsorship. Chasing down thirty-two industry representatives took a lot of effort, but we were most appreciative that they helped take the sting out of the cost of providing a first-class conference.

To paraphrase Arthur T. Hadley, thirteenth President of Yale University (1899-1921), and referenced numerous times by researchers since, if there is any perceived taint to receiving funds from private industry for the advancement of science, it clearly t'aint enough![169] Private/public partnerships are established in all walks of life. When conducted in a transparent manner, it generates a win-win for all.

> The OMAFRA restructuring was good for OSCIA and we continue to strengthen the bond between OMAFRA and OSCIA for objective, non-biased, on-farm research

Where is OSCIA Headed with On-Farm Research?

OSCIA was formed in 1939 under the leadership of innovative farmer, Alex M. Stewart, because extension staff from the department of agriculture and scientists at the Ontario Agriculture College wanted to collaborate with a network of farmers who were interested in new plant varieties. There was also a need to communicate the latest research results. What better way to organize a network of farmers across Ontario than to structure local associations based on counties/districts? Extension workers and scientists could work with this network as on-farm research co-operators who could provide the practical experience. In those early days, the department's county/district agricultural representatives played a key role in organizing on-farm trials and promoting field days and crop tours. A sampling of OSCIA's research projects from the OSCIA communication package of 2006 is outlined in Appendix 10. Each year, a communication package outlining similar projects are circulated to each provincial director, who in turn present these details to each county/district at their local annual meetings.

---

169   Russell, H.L. (1931) "Commercial support for agricultural research". Proceedings of the Forty-Fifth Annual Convention of the Association of Land Grant Colleges and Universities. Burlington, VT: Free Press

Agricultural representatives played an active role until 1999, when Ontario's agriculture ministry went through massive restructuring. OMAFRA shifted to a regional concept that deployed provincial leads with specialties in areas such as cereal crops or horticulture or livestock, a structure that still exists today. The collaboration with university scientists is more challenging today, but the relationship with OMAFRA specialists is much greater for on-farm research. The OMAFRA restructuring was good for OSCIA and we continue to strengthen the bond between OMAFRA and OSCIA for objective, non-biased, on-farm research. An up-to-date listing of OSCIA's on-farm research is available on the OSCIA website at Crop Advances.[170]

Here are more examples of current research projects supported by OMAFRA for OSCIA involving applied, on-farm field activities:

- **Tier One Grants:** exclusively available to local/regional associations to support educational activities, field days, guest speakers, bus tours, in-field trials, or demonstration of new equipment or management techniques. These grants supported forty-one projects in 2016 with $30,000 from the ministry and OSCIA's provincial office contributing an additional $10,000. Local OSCIA associations added more than $53,000 in cash and $35,000 worth of in kind support for the projects.

- **Tier Two Grants:** support applied research projects proposed through local/regional associations and selected through a competitive merit-based process. The three-year projects develop, validate, and demonstrate innovative technologies or new best management practices. Up to $46,000 per year is provided through the ministry, while OSCIA and other sources contribute $54,000. The following on-farm research investigations were supported between 2015 and 2018, carried out in co-operation with OMAFRA experts, academics, local/regional OSCIA members, and industry experts:

    ○ **Roots Not Iron** was a project of the Thames Valley Region SCIA and compares standard cover crop practices to the

---

170   Crop Advances. www.ontariosoilcrop.org/research-resources/crop-advances/ (Accessed June 19, 2017).

concept of year-round green cover and will assess impacts on yield, profitability, soil health, and nutrient availability across all treatments.

- **Advancing Cover Crop Systems in Ontario** focused on soil nutrients, soil health, insects, and nematodes and was co-ordinated through the St. Clair Region SCIA. The project compared cover crop mixes to single species, measured the impact of cover crops on phosphorus loss, and provided information to producers on how to grow cover crops successfully.

- **Environmentally Sustainable Utilization of N in Corn** investigated the yield response of late-season nitrogen applications using Y-Drop technology. Project was led by the Ottawa-Rideau Region.

- **Rapid Development of Farmland from Boreal Forest** helped farmers in the Cochrane and Temiskaming Districts bring brush and lightly treed forested land into agricultural production as quickly as possible using new clearing and cropping techniques. [171]

In 2014, through discussions with then-OAC dean Rob Gordon, we pondered the challenges university scientists face in developing practical outcomes for their research. After all, tens of millions of research dollars were being invested and it was critical that funding would lead to practical benefits. I inquired, and the OSCIA board agreed for me to change my role and job description to work closer with key scientists on outreach. Since OSCIA's number one research priority is to improve soil health and soil quality, Dr. Gordon agreed to provide me with a part-time office in the School of Environmental Science, close to the soil scientists. Coincidently, by 2015, OMAFRA also identified soil health as one of its highest priorities. There were several reasons for this. Healthy soils:

---

171   OSCIA internal files. (Accessed June 14, 2017).

- Provide minimum resistance to root growth and improve crop development, yields, and quality;

- Provide better returns on crop inputs such as applied nutrients and pesticides;

- Allow for better infiltration, more water storage, and less runoff;

- Are more resilient during low water conditions because their structure and organic matter content help retain plant-available moisture;

- Resist degradation, such as compaction, crusting, water and wind erosion, and ponding;

- Are better equipped to remove pollutants and protect groundwater quality; and

- Reduce greenhouse gas emissions, i.e. carbon dioxide, methane, nitrous oxide. Implementing BMPs for soil health, especially those that add organic matter, will improve the ability of soils to serve as carbon and nitrogen sinks.[172]

> Fixing soil is a continuous process

Farmers are keen to work with extension staff and scientists to learn more about improving soil health. They truly understand the importance of healthy soils, healthy crops, and healthy food. It is paramount to their livelihood. Time after time throughout my career, I have heard farmers express how they're working to leave the soil for the next generation in a much healthier state than when they took over. This is an ethic held by most family farms. Fixing soil is a continuous process.

However, the makeup of soil is extremely complex. Scientists know a great deal about the physical and chemical properties of soil, but little is understood about its biological makeup—the billions of microorganisms that live in a handful of the black earth.

---

172   Best Management Practices Factsheet. Soil Health in Ontario. OMAFRA Soils Team (Service Ontario Publications).

The great news is that now through DNA sequencing and genetic barcoding, these little rascals can be identified and counted. We are just beginning to use this technology as a new tool for soil health. Some of the microbes are good, and we want to learn how to increase their population. Others can be nasty in so many ways, and we want to learn how to diminish their influence. These are indeed exciting times for those establishing careers in soil science.

First of its Kind in North America: Research on Soil Ecosystem Services

It is wonderful to stand on the edge of a new research installation that is also the first of its kind in North America. I had this privilege in 2016, when the University of Guelph established the new Soil Health Interpretive Center (SHIC) at the Elora Research Station. OSCIA has an exciting role to assist with outreach under a formal Knowledge Translation and Transfer (KTT) project.

Figure 12.2 Installation of a soil column. This is one of eighteen such columns with sensors installed at various depths. Located at the Soil Health Interpretive Centre at the University of Guelph's Elora Research Station. (Photo credit: Dr. Claudia Wagner-Riddle)

A key component of SHIC is a $2-million lysimeter installation funded by the Canada Foundation for Innovation, and the Ontario Ministry of Research, Innovation, and Science. A lysimeter can measure moisture and nutrients going into the soil as well as the quantity leaving the soil. The SHIC lysimeter consists of eighteen soil columns that are extensively monitored with hundreds of sensors that will collect data to enhance understanding of the environmental impacts of different cropping systems. The SHIC installation will be especially effective at measuring greenhouse gas emissions from alternative cropping systems.

The key objectives of the SHIC are:

- Increasing awareness of research conducted on soil health and ecosystem services by connecting directly with grassroots stakeholders through OSCIA;
- Translation of research results and data into accessible formats for farmers' use and to generate increased understanding of diverse cropping system benefits and drawbacks;
- Engaging farmers to be a part of the research process to increase willingness for implementation of best management practices such as diverse cropping; and
- Educating agricultural producers, policymakers, and students about soil health and ecosystem services.[173]

Tillage 2000: Advanced On-Farm Research

On-farm research took a leap forward in the mid-1980s with the Tillage 2000 program that involved twenty-three farmer co-operators, many of whom were OSCIA members.

The main objective of Tillage 2000 was to develop and evaluate conservation farming systems that maximized economic productivity and minimized

---

[173] Wagner-Riddle Lab. Soil Health Interpretive Centre (University of Guelph) www.uoguelph.ca/ses/claudiawagnerriddle/soil-health-interpretive-center-shic (Accessed Aug. 30, 2017).

soil degradation for specific soil types. The project was eventually expanded to forty farms across the province.

Tillage 2000 was unique to soil conservation in Ontario by virtue of its methodology and process: the project included both a research component and demonstrations within an economic farm unit framework. The process was both investigative and developmental over several years. The program was designed to introduce concepts of conservation tillage systems to a larger number of producers and to provide a way to distribute information and experience through field-scale demonstration.[174]

Tillage 2000 was a co-operative program led by OMAFRA soil conservation advisers with additional expertise provided by the University of Guelph and farm co-operators from the OSCIA. It laid the groundwork for scientific investigation of the conservation tillage movement across Ontario, illustrating how conservation tillage and planting methods save the soil, maintain yields, and save labour and fuel.

After OMAFRA reshuffled its staff resources in 1999, the roles of conservation advisers were merged into provincial leads as extension staff focused on key specialties such as corn, cereals, soybeans, dry edible beans, canola, soil conservation, nutrient management, horticultural crops, and more. It happened in livestock and poultry too. These extension leads are among the best in the world.[175] A full list of the OMAFRA specialists can be found on the OMAFRA website.

Crop Advances

OSCIA adopted its regional structure in parallel with OMAFRA's reorganization in 1999. This established a partnership for each OSCIA region that matched an OMAFRA field crop specialist as a support person for OSCIA.

---

174   SWEEP. Tillage 2000. www.agrienvarchive.ca/sweep/int2000.html (Accessed June 8, 2017).
175   A full list of specialists can be found on the OMAFRA website. See Resources. OMAFRA. www.omafra.gov.on.ca/english/crops/index.html#field (Accessed June 8, 2017).

This has been an excellent working model for applied research. OMAFRA technical experts collaborate with OSCIA farmers for applied on-farm research projects and provide technical expertise to our local and regional boards while at the same time carrying out their responsibilities as a provincial lead specialist.

> On-farm research is at the core of OSCIA members' interests as it generally relates to production and profitability

The Field Crop Unit at OMAFRA established a defined protocol for on-farm applied research. Each year, the results of these many trials are published in *Crop Advances*, a monumental task assembled by Dr. Ian McDonald. Issues of *Crop Advances* can be found on OSCIA's website dating from 2003 to the present.[176]

One of OSCIA's strengths has been our ability to conduct on-farm research and carry out field days, bus tours, twilight gatherings, and workshops to kick the tires of new technology. This would not be possible if not for the support of the OMAFRA Field Crop Unit and modest seed money to pay expenses unique to these trials. Often, there are laboratory costs, extra labour, equipment rental, and signage to promote the trials. There are also substantial contributions made by our local associations in cash or in kind; those contributions amounted to $224,092 in 2015 and $115,415 in 2016. Farmers contribute their time, equipment, and crop inputs. Local farm supply businesses are generous in also donating to the cause with crop inputs. Additionally, local businesses frequently spring for coffee, donuts, or even a barbecue, often in the equipment shed of a loyal OSCIA member. On-farm research is at the core of OSCIA members' interests as it generally relates to production and profitability. Profit is an essential component of farming, as it provides assurance that the bills will get paid, an essential component for sustainable farming.

One of our ongoing challenges has been setting up an online database to do a better job of tracking these trials, conduct analyses, and profile results. Beyond Ontario's borders, the Iowa On-Farm Network is a model recently

---

176   Crop Advances, https://www.ontariosoilcrop.org/research-resources/crop-advances/ (Accessed January 10, 2018).

investigated.¹⁷⁷ We invited their operations manager, Tristan Mueller, to present at a workshop in Guelph, Ont., in June 2016. Other on-farm research data management models have been explored in Illinois and Nebraska. The seed has been sown for Ontario to elevate the *Crop Advances* reports to the next generation of data management. Could field data from precision agriculture be the next generation of on-farm trials? Many expert industry reps, extension staff, and academic institutions are working on an efficient way to collect, analyze, and report on the massive amount of data being collected by GPS systems used in precision agriculture.

> As humans, our daily consumption of food is estimated to include over 100 trillion genes

Sorting Seeds for the Twenty-first Century

It is impossible to talk about vegetable shapes, sizes, and colour, crop varieties, breeds of cattle, or human characteristics without discussing genes or genetic makeup. As humans, our daily consumption of food is estimated to include over 100 trillion genes!¹⁷⁸

To some extent, we are what we eat. Does one modified gene out of the 100 trillion eaten per person per day matter? If my food has been approved for eating by over two hundred toxicologists at Health Canada and it's legal in this country, that's good enough for me. Health Canada is the best in the world as their screening process is highly technical, science-based, and credible.¹⁷⁹

---

177   Iowa Soybean Association, On-Farm Network. www.isafarmnet.com (Accessed June 8, 2017).

178   Anthes, E. "Dining on DNA." Wonderland. PLOS Blogs. (Feb. 9, 2011) www.blogs.plos.org/wonderland/2011/02/09/dining-on-dna/ (Accessed June 27, 2017).

179   Frequently Asked Questions – Biotechnology and Genetically Modified Foods. Health Canada. https://www.canada.ca/en/health-canada/services/food-nutrition/genetically-modified-foods-other-novel-foods/factsheets-frequently-asked-questions/part-1-regulation-novel-foods.html (Accessed July 6, 2017).

Our progressive farmer members work with experts to obtain unbiased results when testing new methods or the traits of new crop varieties. There is no question among the OSCIA members that the science and technology invested in genetic improvement has solved and will continue to solve problems in agricultural production. Genetic improvements and new traits for cover crops will continue to progress as a valuable tool for improved soil management.

For the non-farming public, there is little interest in how we solve agronomic problems but there is keen interest in health and nutrition of food. Genetic improvements through various plant breeding techniques are tackling health, safety and nutritional needs too. Nutrient rich foods (such as Golden rice with enhanced vitamin A), edible vaccines, and genetically engineered plants and microorganisms that biodegrade pollutants are just a few examples. [180]

A meta-analysis of the impacts of genetic engineered crops published by Wilhelm Klümper and Matin Qaim in 2014 reported that the adoption of genetic engineering technology reduced pesticide applications by 36.93 per cent, increased crop yields by 21.57 per cent, while improving farmer profits by 68.21 per cent.[181]

Let your imagination run wild to anticipate how, over the next thirty years, biotechnology will help our soil fixers in ways not yet invented.

I have had the privilege of working as part-time secretary manager for the Ontario Seed Growers' Association (OSGA) since 2003. Perhaps this has given me a different perspective on the world of plant breeding, genetic traits, crop improvement, and the overall benefits to the industry and consumers. Today, any new genetic trait receives intensive screening and testing before

---

180 Afzal, H., Zahid, K., Ali, Q., Sarwa, K., Shakoor, S. et al. (2016) "Role of Biotechnology in Improving Human Health." Journal of Molecular Biomarkers & Diagnosis 8:309 (2016) www.omicsonline.org/open-access/role-of-biotechnology-in-improving-human-health-2155-9929-1000309.php?aid=82443 (Accessed July 6, 2017).

181 Klumper, W. and Qaim, M. "A Meta-Analysis of the Impacts of Genetically Modified Crops." PLOS One. www.journals.plos.org/plosone/article?id=10.1371/journal.pone.0111629. (Accessed July 23, 2017).

being approved for release. This provides me with a high level of assurances regarding the health and safety of these products.

My position with the OSGA has been integrated into OSCIA management duties (for a small fee) under a long-standing informal agreement to combine staff resources. My OSGA duties began when I took over the role of executive director for OSCIA in 2003. OSGA has a separate board of directors and is governed under a separate set of bylaws. Working with the seed industry is simply an extension of OSCIA's mission to facilitate responsible economic management of soil, water, air and crops through development and communication of innovative farming practices.

> Seed certification in Canada works like a well-oiled machine

OSGA has provided me an opportunity to travel across Canada to learn about seed and crop developments in other jurisdictions. Provincial branches of seed growers, organized under the auspices of the Canadian Seed Growers' Association (CSGA), assemble every summer for their annual conference, hosted by one of the provincial branches. Aside from association business and keynote speakers, highlights of the conferences include tours of significant industry establishments or other notable excursions organized by the hosts.

Work with the seed industry for the past fifteen years has allowed me to gain huge insight and understanding of the importance of certified seed and the rigour of the regulations under the Seeds Act. Seed certification in Canada works like a well-oiled machine to protect the buyers, consumers, and of course, the growers. I have been witness to circumstances where if one of the links in the regulatory process was broken (whether intentionally or unintentionally), inferior quality resulted. I have had farmers tell me that imported seed did not always measure up to the high standards of Canadian certification. Weed seed contamination or off-types due to genetic impurity are the most common challenges for seed growers.

My first national seed growers' convention, hosted by the Ontario branch, was held in downtown Ottawa in 2004. This happened to be the hundredth anniversary of the CSGA, so the conference was titled Seeding the Next Century. What a party! It was, by far, the largest conference with which I had been involved as an organizer.

In November 2003, I was thrown into the logistics to help plan for the hundredth anniversary celebration to be held the first week of July 2004. Fortunately, the event co-chairs, OSGA president Bob Hart and Bill Ingratta, director of the Crop Technology Branch, Agriculture and Rural Division, OMAF, had the planning well in hand. Guests from across Canada enjoyed numerous family activities including rafting on the Ottawa River, buses to the Canadian Aviation Museum for the family barbecue, tours of the Ottawa Experimental Farm, the president's reception at the Museum of Civilization, and a bus tour into farm country. Of course, seed business was conducted, too, by the seed grower delegates in the downtown Ottawa hotel.

> After fourteen years, we cannot verify that they'd be waffable

There was an anniversary cake that cost close to $500. We had a string trio for the banquet reception, a piper to accompany the head table guests to their seats, and a vocalist for the national anthem (the singer was the father of the piper and both were Ottawa police officers). Oh, and the wine! SeCan (the largest selling brand of certified seed in Canada) prepared a special hundredth anniversary label for the chosen vintage. I still have several bottles on the wine rack in my basement. Peter Szentimrey, OSGA's 2018 president, indicates he's squirrelled a few bottles away too. After fourteen years, we cannot verify that they'd be waffable. The host team of organizers had special hats, pins, and shirts that identified them as go-to resource people for our guests. The national office, and especially Gaye O'Bertos, executive assistant for the CSGA, has been at the heart of making each annual meeting a success.

Each annual meeting of the CSGA hosted by one of the provincial branches brought unique memories. A few years after I took on my administrative duties with the OSGA, the Alberta branch hosted delegates from across Canada in Edmonton. Traditionally, on the Saturday following the conference, the local branch organizes a tour taking in local attractions and farm innovation. Showing Alberta hospitality at its finest, the locals organized a day trip to tour the oil sands near Fort McMurray. After a short flight from Edmonton aboard three private planes (at our own expense), a full busload was assembled at Fort McMurray for a full-day guided tour that took us right down into the bottom of the quarry, where we learned first-hand the nuances of oil extraction from the tar sands. The infrastructure was massive. The farmers were all envious of the 400-tonne payload of the quarry trucks, surmising how few trips it would take to send their grain to market, compared to a twenty to forty tonne payload of typical farm trucks.

> Certified seed provides a small royalty back to the plant breeder to support reinvestment in future improved varieties

Verifiable Quality

The OSGA, through the CSGA, does not register vegetable seed production. Rather, vegetable certification, although still certified under the Seeds Act, follows a different path for certification through the Canadian Food Inspection Agency (CFIA). Having grown potatoes on our family farm for a number of years, I have seen first-hand the importance of seed certification and its enforcement. Potatoes, perhaps more than any other crop, are susceptible to a long list of diseases and bacterial infections that can devastate a crop. To an unsuspecting buyer, a perfect looking potato planted as seed can be a carrier of nasty viruses. It's important to start with clean, pure stock traceable from the plant breeder and vetted by a certification process tracking that variety's development. This should include best practices for sanitation and antiseptic procedures by the seed grower, as well as inspection under the authority of the CFIA.

Farmer-saved seed has been a controversial topic. This refers to situations where a farmer may retain seed in a bin from last year's harvest, a move that might seem logical to non-farmers but which does not follow the certified seed process. In some circumstances, it may prove disastrous. As an example, I came across a circumstance where an organic grower bought uncertified potato seed from another organic grower because they placed a high priority on seed being organically grown, rather than being certified under the regulatory system. There was no inspection or due diligence to follow proper protocol. The subsequent potato crop failed because of transfer of bacterial rot. It wasn't visible to the eye at planting but it sure took over in the field, destroying their potato crop!

Seed regulations permit farmers to save seed for their own use if that specific variety does not have legal restrictions put in place by the developer/seller. For cereal grains or soybeans, for example, one should take precautions similar to a trained seed grower for farmer saved seed: testing for adequate germination, maintaining weed-free fields, cleaning of weed seeds and debris from the harvested seed, and roguing (physically inspecting and removing undesirable plants) within the field before harvest to remove off-types. I would not grow my own farmer-saved seed without sending a representative sample to a certified laboratory to verify germination, weed free status, purity, and vigour. Vigour of farmer-saved seed generally declines after several harvests, which may affect yield or productivity. Almost all field crops in Ontario, with exceptions of some cereals, are grown from certified seeds. More and more, the marketplace demands it. Further, certified seed provides a small royalty back to the plant breeder to support reinvestment in future improved varieties, regardless of the breeding methods used.

The seed industry in Canada is highly regulated under the Seeds Act, a portion of which is administered by the CSGA. The formation of CSGA, the Seeds Act and many subsequent regulations, and close supervision by CFIA serve a common purpose:

- To ensure, and certify to, the varietal purity of seed produced by its members and to maintain the pedigree thereof;

- To identify, and certify to, for purposes other than further pedigreeing, the varietal purity of seed produced from superior propagating material;

- To encourage the development and introduction of superior varieties and strains of plants;

- To develop programs which expand the use of pedigreed seed for increased domestic crop production and for export;

- To contribute to the establishment and maintenance of high standards in yield and quality of agriculture crops;

- To co-operate with other agencies which have an interest in seed production, promotion and distribution both in Canada and abroad; and

- To co-ordinate the endeavours of pedigreed seed growers with those of plant breeders and crop producers in general.[182]

Did you ever want to learn about plant genetics? An eye-opening visit for me was the tour of Agriculture and Agri-Food Canada's Research and Development Centre in Saskatoon, Sask. There are twenty such centres across Canada but the headquarters is on the campus of the University of Saskatchewan. There you'll find the accumulated knowledge of literally tens of thousands of field research trials of cereals, oilseeds, and legumes.[183]

> I am assured that Canada is doing its part to encourage plant diversity with its many plant breeders working out of a multitude of institutions

Additionally, Plant Gene Resources Canada (PGRC) is centred there with over 110,000 seed samples preserved in the gene bank. "The mandate of PGRC is to acquire, preserve and evaluate the genetic diversity of crops and their wild relatives with focus on germplasm of economic importance or potential for Canada."[184]

In June 2017, OSGA organized a field day at the Elora Research Station to observe cereal crop trials. We were particularly interested in winter wheat varieties being tested under the supervision of Dr. Ali Navabi, a University of Guelph cereal plant breeder. Dr. Navabi and his team were studying 3,700 varieties and 18,000 plots, which included testing winter wheat varieties

---

182   Canadian Seed Growers' Association. Mission Statement and Objectives. www.seedgrowers.ca/about-csga/mission-statement-objectives/ (Accessed June 14, 2017).

183   Saskatoon Research and Development Centre. www.agr.gc.ca/eng/science-and-innovation/research-centres/saskatchewan/saskatoon-research-and-development-centre/?id=1180626618960 (Accessed June 14, 2017).

184   About Plant Gene Resource Canada. www.pgrc3.agr.gc.ca/about-propos_e.html (Accessed on June 14, 2017).

from Nepal for drought resistance. A major focus of Dr. Navabi's research is to develop varieties that have resilience or resistance to fusarium head blight, a devastating fungal disease that infects the grain to produce a mycotoxin called Deoxynivalenol (DON).[185]

While some have expressed concern about the loss of genetic diversity as our food system evolves into a narrower band of genetic material owned by a few multinationals, I am assured that Canada is doing its part to encourage plant diversity with its many plant breeders working out of a multitude of institutions.

Winter wheat is incredibly important for farmers to include in their crop rotation because it provides soil cover throughout the winter and early spring when fields are most vulnerable to soil erosion. Winter wheat is a soil protector and soil fixer! It also works well when farmers apply red clover in the early spring to germinate as a cover crop after wheat harvest. Red clover is also a soil builder and legume, which adds nitrogen to the soil. Red clover generally flourishes as a cover crop after wheat harvest, plus it has the ability to grow abundantly into the late fall and overwinter into the following spring, providing another winter of soil protection. The dedication of plant breeders and other scientists at the multitude of research facilities around the world will generate some dramatic improvements in staples such as wheat, rice, potatoes, and others for generations to come.

Plant breeding and genetics is a universe unto itself. Those working within this universe are highly educated, skilled and patient. Improvement of just one genetic trait traditionally takes ten years, if not a lifetime, and development typically costs millions of dollars. To better understand the complexity of plant breeding, see the chart below developed by Dr. Kevin Folta, professor and chairman of the Horticultural Sciences Department at the University of Florida, Gainesville.

---

185   Sobrova, P., Vojtech, A., Vasatkova, A., Beklova, M., Zeman, L., and Kizek, R. "Deoxynivalenol and its toxicity." PubMed Central®. U.S. National Institutes of Health's National Library of Medicine. www.ncbi.nlm.nih.gov/pmc/articles/PMC2984136/

Plant Breeding Techniques

	Traditional Breeding	Hybrids (Crossed inbreds)	Polyploids	Mutation Breeding	Interspecific Crosses	Transgenic	Cisgenic /intragenic	Gene Editing
Examples in Common Foods	Many crops	Corn	Strawberry wheat bananas	Barley, citrus, pears, apples, yam	Pluots, tangelos	Corn, soya, canola, sugar, beet, papaya	Potato, apple	Many coming
Genes combines across species	Sometimes	Sometimes	Yes, often	No	By definition	Yes	No	No
Occurs in Nature	Yes	Yes	Yes	Yes, transposon movement, spontaneous mutation	Yes rare, not always fertile	Yes, agrobacterium, other examples	N/A	N/A
Human Intervention	Yes, for crop improvement	Yes, human facilitated	Can be chemically induced	Yes, to create variation in traits	Yes, for crop improvement	Yes, for precise addition or subtraction of a gene	Yes, for precision crop improvement	Yes, for precise changes in DNA
Number of genes affected*	10K to >100K	10K to >100K	Could be >100K	No way to easily assess	10K to 100K	1-3	1	1
Know what genes affected do	No	No	No	No	No	Yes	Yes	Yes
Plant Patentable	Yes	Yes	Yes	Yes	Yes	Yes	Yes	Yes
Documented adverse effects	Yes	Yes	???	???	Yes	No	No	No
Environmental Assessments	No	No	No	No	No	Yes	Yes	TBD
Organic Acceptable	Yes	Yes	Yes	Yes	Yes	No	No	No
Time for new variety	5-50 years	5 years	>5 years	>5 years	>5 years	<5 years†	<5 years†	<5 years†
Demanding Labels?	No	No	No	No	No	Yes	Yes	??

*The number of genes affected includes allelic variants, paralogs and homoeologs, which are common in plants. †Product development, not regulatory assessment

Figure: 12.3 Comparison of Crop Genetic Improvement Techniques.[186]

I presented an earlier version of Dr. Folta's chart to delegates at OSCIA's annual meeting several years ago. For many, the chart cleared up confusion and misunderstandings about plant breeding techniques. "When compared

---

186 Source: Folta, K. Chart provided through personal communication with Dr. Kevin Folta, Professor and Chairman, Horticultural Science Department, University of Florida, January 24, 2018.

to older breeding techniques, the transgenic method is much more precise and controlled," notes Dr. Folta.[187]

I continue to ponder why strong opinions are targeted against genetic engineering by individuals without any understanding of plant breeding techniques, its history, applicability, or the science behind it. All foods we eat have been developed by genetic modifications in one way or another. For example, why would food developed by mutation breeding be any safer with unknown effects to many genes than transgenic breeding where we do know the effects on one to three genes? The average consumer cannot help but be confused, with or without GMO labelling on packaging. At minimum, the chart above confirms that a simple label cannot provide a clear picture of the process by which the foods we eat have acquired the traits we know and enjoy.

> Genetic engineering is a process, not an ingredient

For those who are uncertain about various forms of genetic development, ask the practitioners, especially those who manage their crops. Ask the scientists who have a lifetime of experience with plant breeding. Ask extension personnel who are trained in agronomy and science. Ask a nutritionist about our daily dietary requirements. Genetic engineering is a process, not an ingredient.

Contrary to myths circulating in social media, the use of genetically engineered crops in Ontario has not increased herbicide resistance. Research by OMAFRA suggests the opposite is true when compared to some of the older technology.[188]

Gene editing research underway in laboratories around the world is poised to solve a wide range of plant-related, microbial, animal, and human health problems. I anticipate that in thirty years, the next installment of OSCIA's history will outline a plethora of benefits. Canada has already approved the non-browning Arctic apple variety. The Innate potato variety has also been approved to reduce

---

187   Ibid.
188   "Has the adoption of genetically engineered (GE) crops resulted in more herbicide resistant weeds in Ontario?" Field Crop News, OMAFRA (December 2016). www.fieldcropnews.com/2016/12/has-the-adoption-of-genetically-engineered-ge-crops-resulted-in-more-herbicide-resistant-weeds-in-ontario/ (Accessed May 4, 2017).

the carcinogen acrylamide, a chemical by-product caused by cooking at high temperatures associated with fried food. Additionally, the Innate potato reduces bruising by up to 44 per cent, which results in less food waste.[189]

Being from a potato farming family, I can't help but be enthused about the second generation of the Innate potato that resists late blight and eliminates the use of chemical sprays. This blight is the same devastating fungal disease that led to the Irish Potato Famine in the 1840s and continues to threaten potato crops in modern times. Currently, potato blight is controlled with fungicidal sprays. With the tweak of a gene, this pesticide can be mostly eliminated from a farmer's toolkit.[190]

OSCIA members, typical of most farmers in North America, will continue to be soil fixers and problem solvers. The majority of farmed acres in Canada that I have visited are managed by sophisticated business people who carry sharp pencils, and who are always looking for ways to cut costs per unit of production. Driven by a philosophy of continual improvement, farmers seek out superior crop varieties and better ways to manage insects, disease, weeds, and their soil. If there's a way to cut expenses without risking their productivity, they'll do it. They're always fixing what wasn't quite up to par the year before. This is a common characteristic among all farmers, regardless of farming system or farm size, whether mainstream or organic. The majority of consumers trust farmers to produce safe, wholesome food. We are so fortunate in the Western world to have so many choices. Most of the world's population does not.

Weather or Not

Weather! We talk about it constantly and farmers have an especially keen interest because, well, their crops are dependent upon the weather. Further, heavy rainstorms can be devastating to our precious soil.

---

189   "Innate: Less Waste, More Potato." www.innatepotatoes.com/ (Accessed July 21, 2017).

190   "Innate® Second Generation Potato Receives Canadian Government Clearance." www.innatepotatoes.com/newsroom/view-news/innate-second-generation-potato-receives-canadian-government-clearance (Accessed August 18, 2017).

Farmers are always watching the weather: observing the sky before heading indoors in the evening, scanning cloud formations in the morning, hypothesizing about the thickness of the squirrel's coat in the fall, or observing the quantity of acorns on the oak tree. Favourable weather is as an important ingredient as any for farming success.

Success in growing crops depends upon determining when the soil has sufficient moisture, but not too much, for timely planting, germination of seeds, and establishment of a healthy root system.

Being able to predict the weather, especially severe rainfall that can cause devastating soil erosion, would be every farmer's dream. Tools to predict the vagaries of weather have very much been part of research and development over the past thirty years. OSCIA too has a vested interest.

If I were to identify any major disappointments in my thirty-year career, one of them would have been OSCIA's inability to gain funding support to establish an agricultural weather centre (AGWEC) for Ontario. By 1995, Environment Canada had decided to close down many regional weather offices, not only putting numerous highly qualified meteorologists out of work, but also, it was feared, jeopardizing the availability of accurate local weather forecasts. At the time, we were told that their regional weather offices received nearly a million calls per year from the agricultural industry for the latest weather forecast.[191]

The government's regional weather offices fed their forecasts into local radio and TV stations, provided timely recordings on toll-free telephone numbers, and generally predicted how emerging weather patterns would affect their local citizens. The new model of centralizing weather forecasts in major cities across Canada seemed unreasonable and unacceptable to agriculture.

The OSCIA board encouraged me to investigate. After various meetings with a few highly qualified and respected meteorologists, options were discussed. Our OSCIA office began to assemble a proposal that would establish an AGWEC for Ontario that would focus strictly on the needs of the agricultural community. The AGWEC proposal outlined requirements for funding to provide basic, timely, and accurate weather forecast service. In addition, AGWEC would offer

---

191 AGWEC Business Plan, OSCIA internal files.

value-added services, such as providing long-term weather trends and predictive models for disease and insect management. The business plan suggested that through sales and services, AGWEC would be self-financing in five years.

Coincidently, a new funding agency, the Ontario Agricultural Adaptation Council (AAC), was emerging as an independent body that would be entrusted with millions of dollars from Agriculture and Agri-Food Canada (AAFC) to invest in projects to advance the competitiveness of Canadian agriculture. Could this agency be the ticket to provide funding for the establishment of AGWEC?

The AAC board was made up of highly respected farm and industry leaders to review and consider funding for innovative projects. From OSCIA's perspective, it was serendipitous that the AAC funding could be available to meet the needs of a new agricultural weather centre, speculating that reliable weather forecasts would be a priority for the industry. Our application was ambitious and requested a cash infusion of over $1.3 million over four years. But we felt confident that we were moving the industry into a new era of providing ag-focused weather services. Our proposal was one of the first for the newly formed AAC and I'm sure the board members struggled over their new trusteeship and how to best invest public dollars.

Our AGWEC proposal was turned down. Perhaps we were too ambitious in our vision. After all, it had to be self-funding when the AAC funds ran out. The proposal may have appeared too bureaucratic. It didn't help that a few days before the AAC board met that June to rule on our proposal, the forecast over the airwaves predicted clear and sunny weather continuing into the following week. Thousands of acres of hay were cut on the weekend in anticipation of making hay while the sun shines. The forecast was wrong! By Monday, it was raining. I'm sure every AAC board member noticed the incorrect forecast. An unusual inversion had gone undetected and changed the weather pattern quickly and unexpectedly. Thousands of acres of hay were ruined and our proposal was shelved.

All was not lost, however. The watchful eyes of researchers and experts at the University of Guelph Ridgetown Campus were already working on disease forecast models that would gather weather data to predict the likelihood of disease outbreaks in crops. The Ontario Weather Network (OWN) program was launched in 2000 to provide support to farmers through emerging research and new technology. Models were developed to predict potential disease outbreaks based on weather factors, allowing farmers to respond with more timely and efficient fungicide control. An early example was development of a predictive model to combat deoxynivalenol toxin (DON) in wheat. Wheat farmers use a forecasting model called DONcast® to plan timely and effective fungicide applications to protect their crops.

In 2006, that service was privatized, moved to Chatham, and its name changed to Weather Innovations Inc. (WIN). Today, WIN offers numerous services including BEETcast™, SPRAYcast®, BINcast®, SPUDcast™, TOMcast™, TURFcast™, VITIcast™, and WHEATcast™. Additional models are available for apples, corn flea beetle, snap bean leaf beetle, strawberry, and more. Services are delivered to other provinces and countries and now the business manufacturers and distributes sensors and related specialized services. [192]

It has been rewarding to see some of the original ideas of AGWEC now carried out by the WIN organization. When it comes to weather forecasts, these days most farmers including me have several apps on their handheld device that provide accurate forecasts with real-time graphics to helps us track showers or storms travelling across our landscape.

The Organic Paradox

I once got my ear chewed off by an irate caller who was disappointed after attending an Environmental Farm Plan (EFP) workshop. The caller expected that EFP would promote organic farming practices. It became evident that the caller's expectations were not met by any aspect of the workshop agenda, the EFP workbook, or the roundtable discussions where most of the participants were non-organic producers. It also became evident that the caller had urban roots and had recently relocated to a rural property.

In designing the EFP program, the working group of farm leaders and government technical experts carefully avoided promoting one farming

---

192 Weather Innovations. www.weatherinnovations.com/models.cfm (Accessed June 19, 2017).

system over another. Each farmer has a right to develop a unique business plan, commodity choice, market outlet, or farm size. EFP is designed to fit all, and best practices contained in EFP apply to organic and non-organic producers alike.

A few OSCIA members have pursued organic certification on at least a portion of their acreage, but many OSCIA members have adopted no-till as their preferred farming system to eliminate soil erosion. No-till requires herbicides for weed control. Although a few organic growers are experimenting with roller/crimper and green cover techniques, that system is still not proven and carries with it a number of challenges and knowledge gaps related to consistency of weed control, insects, and determining best practices.[193]

The destruction of soil organisms through tillage and leaving bare soil exposed to the elements is contrary to progress our farmers have made over the past thirty years. Degraded soil can be repaired with livestock grazing or dairy farms growing lots of hay and alfalfa while following best practices for manure application. This provides nutrients for plants and stimulates soil microbes—great for soil health whether organic or non-organic. What about field crops of corn, soybeans, and wheat in a rotation with cover crops for green manure? There are many theories and some myths too. Like so many practices on the farm, nothing is black or white. While OSCIA does not promote one farming system over another, it does promote the use of science and supports researchers in their quest to provide answers to farming challenges.

In 2014, following a meeting related to pollinator health near Queens Park in Toronto, I was quizzed by a senior government official about why more farmers were not adopting organic practices. After all, he suggested, the price premiums were significant. Higher prices should motivate farmers to pursue the financial rewards of going organic. In this non-farmer's mind, there could be more profit made with an organic system. In response, I pointed out a

---

193   Vincent-Caboud, L. et al. "Overview of Organic Cover Crop-Based No-Tillage Techniques in Europe: Farmers' Practices and Research Challenges." Agriculture Open Access Journal (2017, 7, 42) www.mdpi.com/2077-0472/7/5/42/pdf (Accessed August 31, 2017).

number of risk factors that are often not apparent to the non-farming public. Those who are well read do understand there is a paradox in the industry—the organic paradox. Why is the apparent strong demand for organic food by consumers (and retailers) not being met by farmers? Demand is so strong that a substantial amount of organic grain and soybeans must be imported.[194]

According to the 2016 Census of Agriculture, 2.2 per cent (4,289) of all farms in Canada are certified organic or transitional to organic.[195]

Before pondering the reasons for farmers' reluctance to adopt organic practices, it is important to clarify what is meant by organic production. Definitions vary from country to country. Organic standards from importing countries do not necessarily meet our Canadian standards.

Canada has laid out a regime to ensure compliance both within Canada and for imported products, including regulations that outline:

- Organic Production Systems General Principles and Management Standards;
- Organic Production Systems, Permitted Substances List;
- Conformity Assessments - General Requirements for Accreditation Bodies; and
- General Requirements for Bodies Operating Product Certification Systems.

In Canada, 'organic' refers to "an agricultural product that has been certified as organic in accordance with the Organic Product Regulations or that has been recognized as such under section 29 of the Regulations" and certified

---

[194] "CoBank Report Shows Rising Demand For Organic and Non-GMO Grains Outpaces U.S. Production." CoBank.com (Feb. 13, 2017) (https://www.prnewswire.com/news-releases/rising-demand-for-organic-and-non-gmo-grains-outpaces-us-production-300406463.html%20Releases/News%20Releases%202017/Organic_NonGMO%20Demand_Press%20Release.pdf, (Accessed July 29, 2017).

[195] "Growing opportunity through innovation in agriculture." Farm and Farm Operator Data, Statistics Canada. www.statcan.gc.ca/pub/95-640-x/2016001/article/14816-eng.htm. (Accessed July 25, 2016).

through a "procedure whereby a CFIA-accredited certification body provides written assurance that agricultural products are organic as defined in and for the purposes of the Regulations. Certification of products may be based on a range of inspection activities including verification of management practices, auditing of quality assurance systems, and in/out production balances."[196]

Barriers to Organic Adoption

1. Crop Yield and Quality

Let's start with crop yield. In Ontario, the crop insurance agency, Agricorp, provides documented average yields that compare individual organic farms to conventional farms. From 2006 to 2014, organic soybean yields ranged from 64.4 per cent to 74 per cent of conventional. Organic corn yields (2011-2015) averaged slightly more than half of conventional yields. Organic wheat (2008-2014) ranged from 47.8 per cent to 66 per cent of conventional.[197]

For example, wheat is grown by most OSCIA members. To ensure highest quality and safety for consumers, most conventional wheat growers apply fungicides to reduce or eliminate fungal diseases. "Until a variety resistant to *fusarium* is found, wheat growers simply must spray a *Fusarium* fungicide," says Peter Johnson, also known as Wheat Pete, formerly the cereals specialist with OMAFRA. "Winter wheat is incredibly important in a crop rotation, both for soil health and to provide soil protection from the forces of nature over the winter months. Our commercial growers understand the importance of producing high quality wheat that is absolutely safe for consumers. A fungicide helps to ensure this."[198]

---

196   Canadian Organic Office Operating Manual. Canadian Food Inspection Agency. www.inspection.gc.ca/food/organic-products/certification-and-verification/operating-manual/eng/1389199079075/1389199224543?chap=2. (Accessed July 24, 2017).

197   Personal communication with Agricorp staff and Harold Rudy (June 27, 2017).

198   Personal communication with Peter Johnson and Harold Rudy (Aug. 2, 2017).

For horticultural producers who grow the fruits and vegetables seen on grocery store shelves or at farmers' markets, disease and insects are the greatest threat to yield and quality. Weeds can also be a threat but with strategic planning, available labour supply, a bit of luck from the weather and wise use of approved management tools and extensive tillage, weed control can be mastered by organic growers. Farmers tell me that to ensure timely field operations for weed control, the soil must be exceptionally well drained, generally with artificial drains. The benefits of no-till versus conventional tillage for soil health were already discussed in earlier chapters.

2. Price Insecurity

If organic crop yields are reduced by between one-third to one-half, what about the price premiums? Organic products generally command a premium price. A global meta-analysis conducted by Crowder et al. suggests that organic agriculture can be up to 35 per cent more profitable. However, the study was a literature review of studies in locations where the estimated yield reduction in the organic systems was only 10 to 18 per cent lower than conventional, not typical of the 25 to 50 per cent reduction that Agricorp has found in Ontario.[199]

In addition, prices for organic products have been volatile and less predictable than for conventional crops. Furthermore, especially for horticultural produce, studies show the large grocery chains have become more aggressive at driving prices down closer to conventional prices.[200]

Based on discussions with Ontario organic growers, the wholesalers and retailers are driving prices lower for organic products in Ontario as well. It is becoming more challenging for organic producers faced with rising costs of production while the prices being offered for organic produce are declining.

---

199   Crowder, D.W. and Reganold, J.P. "Financial competitiveness of organic agriculture on a global scale." Proceedings of the National Academy of Sciences (PNAS) Open Access. www.pnas.org/content/112/24/7611.abstract (Accessed June 22, 2017).

200   Conlin, C. "The Organic Food Industry: An Analysis of Supply and Demand via Aggregate Prices." Penn Libraries, University of Pennsylvania. www.repository.upenn.edu/cgi/viewcontent.cgi?article=1000&context=spur/ (Accessed Aug. 3, 2017).

Fundamental economics would suggest that such a cost-price squeeze will either drive more efficiency through technological advancements, or the business will not be sustainable.

3. Lag Time, Complexity and Expenses for Organic Certification

The regulations for organic certification are complex. A field cannot grow a crop with prohibited organic crop inputs (e.g. synthetic fertilizers and some pesticides) for a period of three years prior to organic certification. This interim lag time is difficult for many growers as crop yields will suffer through the transition before they can enjoy the benefit of premium prices.

For small-scale growers, typically selling direct to consumers through member-based Community Shared Agriculture (CSA) arrangements, "the expense and effort of certification isn't justifiable, since attaining certified organic status doesn't significantly increase sales or the trust already earned from buyers."[201]

4. Aligning with Science for New Tools to Advance the Organic Industry

Farmers are continually working to solve problems and manage risk. I have spoken to leading farmers about the risk of mycotoxins from fusarium in wheat grown without the use of fungicides. Fusarium head blight contains mycotoxins that are harmful to humans and animals.[202]

Fungicides, used in combination with selective breeding for disease-resistant traits, are the farmers' ammunition. What if wheat genes could be edited or modified to completely resist fusarium and eliminate mycotoxins without the use of fungicide sprays? Would the organic industry accept such a technique? What about the use of science by plant breeders to solve plant disease problems in other commodities? How will science versus tradition dictate a path forward?

---

201   "To Certify or Not to Certify: The perspective of small-scale organic farmers." Organic Federation of Canada. www.ofcfbc.wordpress.com/2014/10/30/to-certify-or-not-to-certify-the-perspective-of-small-scale-organic-farmers/ (Accessed Aug. 31, 2015).

202   Zain, Mohamed E. "Impact of mycotoxins on humans and animals." Science Direct, Journal of Saudi Chemical Society. www.sciencedirect.com/science/article/pii/S1319610310000827 (Accessed July 25, 2017).

A recent development where science has created a new opportunity for the organic industry is the approval of digestate for use as fertilizer. Digestate is a fermented, nutrient-rich by-product of anaerobic digestion. Companies like Bio-En Power Inc. of Elmira, Ont., are tapping into improved biodigester technology that uses microorganisms to break down organic materials such as livestock manure and food by-products into methane, which can be used to generate electricity and a natural fertilizer. Digestate from these off-site facilities is exactly what organic farmers need if they don't have access to animal manures, since organic producers are not allowed to use most commercial fertilizers. This product has been approved by the CFIA for organic certification of their crops.

Christine Brown, a nutrient management specialist at OMAFRA, has been investigating the benefits of digestate in collaboration with Gord Green and Stuart Wright (OSCIA's 2016 president and 2018 vice-president respectively) and other researchers. When applied just prior to summer or early fall cover crops, digestate provides a low-cost option that boosts nutrients and enhances plant and root biomass, stimulates microbial activities and builds the soil's resiliency. Research is ongoing to better quantify results.

Scientists, geneticists, and farmers (organic and non-organic) all have the same goal: to make our world a better place by growing healthy, nutritious food while minimizing the environmental impacts and ensuring that farmers are able to make a decent living for their families. One of the most enlightening books I have read that provides a logical perspective on how science can assist the goals of organic producers is by Dr. Pamela Ronald, a geneticist at University of California, Davis. Her husband is an organic farmer and together they've written *Tomorrow's Table: Organic Farming, Genetics, and the Future of Food*.[203] If you're looking for an enlightening way to spend twenty minutes, Dr. Ronald's recent TED talk provides an excellent overview of how the science of genetics can and will provide food security. Better yet, read their book. Enlightened organic producers may wish to take notice.

---

203   Ronald, P.C. and Adamchak, R. W., "Tomorrow's Table: Organic Farming, Genetics and the Future of Food." (2018 edition) https://global.oup.com/academic/product/tomorrows-table-9780199342082?cc=us&lang=en& (Accessed January 17, 2018).

My commentary on why farmers are reluctant to adopt organic production is in no way intended to be critical of an organic farming system. Indeed, there are also many barriers and challenges on non-organic farms. The adage 'better the devil you know than the devil you don't know' is most apropos when it comes to changing farming systems! Some organic farms are doing very well. I have family friends who have been part of Community Shared Agriculture (CSA), where the produce is fresh, locally grown, and serves a loyal client base that shares their values.

> The adage 'better the devil you know than the devil you don't know' is most apropos when it comes to changing farming systems

How have I lasted for over thirty years working at OSCIA with our diversity of membership? I support OSCIA's philosophy of being pro-choice when it comes to farmers choosing a farming system that best meets their needs. I'm also pro-farmer. I embrace a farmer's success rather than passing judgment on their commodity choice, size of operation, marketing strategy, business structure, or farming system. I don't have strong opinions about how farmers go about their business, but I support OSCIA's basic principle that soil and crop management must be based on sound science. Our Canadian consumer is indeed fortunate to have so many food choices.

The Dose Makes the Poison - Chemicals and Food

Recently, I read an article about a respected health practitioner who promoted not just eating an apple a day, but eating apples in their entirety—core, seeds, and all. I thought seeds contained cyanide, a nasty poison with a long sordid history. The health practitioner acknowledged cyanide's lethal attributes in apple seeds but also suggested that there was not enough in them to be of concern. In other words, the dose makes the poison.

In the early 1500s, a Swiss doctor named Philippas Aureolus Theophrastus Bombastus von Hohenheim-Paracelus wrote that "All substances are poisons; there is none which is not poison. The right dose differentiates a poison from a remedy."[204] In modern times, Dr. Joe Schwarcz, a chemist and professor at McGill University, as well as a well-known media commentator, has confirmed similar statements. I have heard Dr. Schwarcz numerous times as a keynote speaker. He is a skilled communicator able to clarify misconceptions about chemicals, food safety, and public concerns about these issues.

The study of soil and what we put on it, in it, or over it is not complete without commenting on chemicals. Chemicals are often perceived as undesirable, but it should be noted that for soil fixers, there is a huge world of chemistry affecting zones around plant roots, microbes, and molecular functions within the plants. But most chemicals in the soil are natural. The study of chemistry, biochemistry, molecular biology, and toxicology are complex topics. Thousands of pages have been written on this subject and it is not my intention to overwhelm the reader with technical facts. Throw in the word "pesticides", however, and most react negatively.

A few chemicals (both natural and synthetic) may be carcinogenic; that is, they can cause cancer. Farmers know this and enrol in workshops to learn and become licensed to purchase and apply pesticides according to strict regulations.

Dr. Schwarcz reports that "70 per cent of fruits and vegetables have no detectible pesticide residues and only about one per cent of the time is the legal limit exceeded, a limit that already has a hundred-fold safety factor built-in."[205]

However, natural pesticides are prevalent in our food too. To put pesticides into perspective, Dr. Bruce Ames and colleagues at University of Berkley, California, estimate that humans may eat 1.5 g of natural pesticides per day, 10,000 times more than the amount synthetic pesticide residues in their food. In fact, Dr. Ames and his team estimate that 99.99 per cent (by weight) of pesticides in our diet are chemicals naturally produced by plants to defend themselves from

---

204 "Assessing Toxic Risk, Student Edition." Cornell University Chapter 1, p.1 http://ei.cornell.edu/teacher/pdf/ATR/ATR_Chapter1_X.pdf (Accessed March 27, 2017).

205 Personal communication with Dr. Joe Schwarcz and Harold Rudy (December 15, 2017).

insects and diseases. They report that only fifty-two naturally occurring pesticides have been studied in animal tests and of these, twenty-seven natural pesticides are known carcinogens. [206]

For soil fixers, this research provides a level of comfort for prudent use of modern-day tools which include pesticides to manage soil. So far, most conservation-minded farmers save their soil from erosion by leaving lots of plant stalks and leaves on the undisturbed soil surface, requiring an herbicide for weed control. Previous centuries and civilizations of agriculture did not have this option and the consequences were devastating. Given the research of Ames et al at Berkley, farmers are reassured that carefully managed herbicides are a tool to protect our soil resources with little risk to our food supply. Organic farmers too are seeking solutions to soil erosion, but the task requires more research and innovation to achieve best practices for soil management.

> Previous centuries and civilizations of agriculture did not have this option and the consequences were devastating

Would you rather eat grain that contains a natural carcinogen called aflatoxin or minimize aflatoxin with a synthetic fungicide with no known health concern? What about selective genetic modification to eliminate unwanted toxins? My choice is to go with best agronomic practices, verified and approved by science. Safe, wholesome food begins with good agronomic practices on the farm. These practices include using a variety of tools to manage weed infestations, insect damage, and plant diseases, including aflatoxins produced naturally from molds that may occur in cereals, oilseeds, spices, and tree nuts under hot, dry conditions and where products are improperly managed. Agricultural pesticides are a tool for good agricultural practices to reduce risk of these molds developing.[207]

---

206 Ames, B.N., Profet, M., Swirsky Gold, L., Dietary pesticides (99.99% all natural). Proc. Nat'l Acad. Sci. USA, Vol. 87, pp. 7777-7781 (October 1990) Medical Sciences

207 "Aflatoxin Impacts and Potential Solutions in Agriculture, Trade, and Health." Partnership for Aflatoxin Control in Africa (PACA). www.un.org/esa/ffd/wp-content/uploads/sites/2/2015/10/PACA_aflatoxin-impacts-paper1.pdf (Accessed July 27, 2017).

Farmers receive extensive training for the selection, purchase, transportation, storage, and application of agricultural pesticides. The Ontario Pesticide Education Program has offered the Grower Pesticide Safety Course for farmers since 1986. Like most Ontario farmers, I attended one of the early workshops to receive my licence. Each participant must pass a test in order to be certified and receive a licence which is mandatory for pesticide purchases. For me, taking this course was enlightening. I gained much more insight into the broad range of chemicals, their use, and the complex process of product registration by the Health Canada's Pest Management Regulatory Agency (PMRA). I also gained great insight into the chemistry, the mode of action, alternatives that could be used, and the importance of chemical rotation to minimize pest resistance. Specific emphasis is on the safe use of products and how to protect yourself, your family, pets and animals, neighbours, the environment, and the food produced.

> Herbicides for weed control have been particularly essential for farmers practicing no-till to protect and improve the soil

In my travels, I've heard the same insight from OSCIA members and other farmers (those who have received the pesticide management training). There is a huge range of toxicity, persistence, efficacy, safety, and effectiveness for a specific targeted species or organism. To not use crop protection products, farmers increase their all-round risk of crop damage due to weeds, insects, or disease. Herbicides for weed control have been particularly essential for farmers practicing no-till to protect and improve the soil.

Organic farmers use chemicals too, but have fewer options. Copper sulphate, for example, is a permitted fungicide under organic systems but I prefer synthetic fungicides because of what I learned at the Ontario Grower Pesticide Safety Course for my certification. Copper sulphate is highly toxic to fish and aquatic invertebrates, as well as bees.[208]

---

208   Extoxnet. Cornell University. www.pmep.cce.cornell.edu/profiles/extoxnet/carbaryl-dicrotophos/copper-sulfate-ext.html (Accessed January 24, 2017).

Not many new active ingredients in pesticides are being developed today, organic or non-organic. According to Jay Bradshaw, president of Syngenta Canada, a product that would have cost $100 million to develop in the year 2000 would cost close to $268 million today. Typically, it takes eleven years from the time a new molecule is discovered to develop a commercial product.[209] Pesticides are expensive. Ontario farmers have learned through Integrated Pest Management (IPM) programs how, where, and when to use sprays only when required. IPM programs reduce the use of chemicals dramatically.

To come full circle on the discussion of chemicals and food—the dose does indeed make the poison—a recent Danish study found the cumulative risk of pesticides in our diet (at least in the diet of Danes) is equivalent to drinking a glass of wine every seven years.[210] How many people would restrict their wine to that amount?

Not everyone accepts the assurances of scientific studies and for personal, ethical, environmental, or cultural reasons, simply do not want pesticides added to crops. What if genetic engineering eliminated threats such as aflatoxin AND eliminated use of a pesticides, providing even greater levels of assurance for food safety? That, in fact, appears to be what's happening with recent developments in gene editing for maize. Crop losses to aflatoxin around the globe are massive and a simple gene silencing technique may have potential to eliminate this threat.[211]

---

[209] Personal communication with Jay Bradshaw and Harold Rudy (Jan. 23, 2018).

[210] Larsson MO, Sloth Nielsen V., Bjerre, N., Laporte, F., Cedergreen, N. "Refined assessment and perspectives on the cumulative risk resulting from the dietary exposure to pesticide residues in the Danish population," Food and Chemical Toxicology. https://www.ncbi.nlm.nih.gov/pubmed/29155356 (Accessed January 8, 2018).

[211] Thakere, D., Zhang J., Wing, R.A., Cotty, P.J., Schmidt, M.A. "Aflatoxin-free transgenic maize using host-induced gene silencing." Science Advances (March 10, 2017). www.advances.sciencemag.org/content/3/3/e1602382 (Accessed August 31, 2017).

Chemicals and the Environment

The use of agricultural chemicals and their potential impacts on our surrounding woodlots, fencerows, ravines, and waterways has been and still is a highly researched topic. For our farmers, the same array of resources for training, management tools, and principles of Integrated Pest Management (IPM) that apply to chemicals and food also come into play with respect to chemicals and the environment. The agricultural industry relies on the federal Pest Management Regulatory Agency (PMRA) under the Pest Control Products Act for approval of products. In consultation with the U.S. Environmental Protection Agency and other international counterparts, hundreds of toxicologists review, test, evaluate (and re-evaluate) the efficacy and safety of each new product.[212]

In Ontario, additional rigour is applied to pesticide use under the Ontario Pesticides Act for training, certification, licencing, sales, transportation, storage, and disposal of pesticides.[213]

Use of pesticides does not always go smoothly. In 2012, some bee producers noted significant bee deaths during spring planting. It was a perfect storm for disaster. Planting weather was warm and ideal to be outdoors, and virtually every corn and soybean planter was in the field within the same week that spring. Conditions were also ideal for the bees to be out seeking spring blossoms.

Surprisingly, and unbeknown to most, talc powder came into play. I don't wear talc but I understand that it provides skin as smooth and slippery as a baby's bum. Well, talc has the same effect in the seed box of a corn planter. Seeds of most annual crops like corn and soybeans were coated in neonicotinoid insecticide designed to keep bugs from eating the crop. Seeds may be sticky as they move through the planter mechanism, so farmers added talc to keep their seed flowing smoothly through the vacuum systems used

---

212 Health Canada. The Regulations of Pesticides in Canada, https://www.canada.ca/en/health-canada/services/consumer-product-safety/reports-publications/pesticides-pest-management/fact-sheets-other-resources/regulation-pesticides.html (Accessed January 12, 2018).

213 OMAFRA. Using Pesticides in Ontario, http://www.omafra.gov.on.ca/english/crops/resource/using-pesticides.htm (Accessed January 12. 2018).

in planters. Unfortunately, the talc contributed to fugitive dust containing neonicotinoid particles drifting into the margins of fields where blossoming trees and plants were attracting bees. The industry required answers on how to better manage exhaust dust from planters.

OSCIA seized the opportunity to become engaged in a unique research project from 2014 to 2017 entitled "Exposure of honey bees to fugitive neonicotinoid-contaminated dusts near corn." Over $1 million was acquired and channeled by OSCIA's office through the Office of Research at the University of Guelph to support the research team, headed up by Dr. Art Schaafsma at the Ridgetown Campus. The Ontario Agricultural Adaptation Council was a significant contributor to this research, but we also received a significant contribution from the U.S.-based Corn Dust Research Consortium, which is a testament to the high-quality expertise in Ontario and the level of interest in finding answers to this problem.

The research resulted in practical recommendations for farmers to ensure that the pesticide stays on the seed as it moves into the soil. This can be achieved by filtering and redirecting the exhaust dust into the soil; attaching a filter on air intakes to remove field dust; incorporate non-abrasive seed lubricants; and using conservation tillage. With these best practices, research proved that chemical residue escapes can be reduced by 90 per cent.[214]

While the research described above got underway, talc powder was banned for agricultural use, neonicotinoids were restricted by implementation of new regulations, and the industry introduced new seed treatment products with more benign environmental impacts.

Gene editing holds much promise for human and environmental health from many perspectives. New developments will eliminate the use of pesticides. The University of Florida's Dr. Kevin Folta provides great insight on genome editing and its possibilities to solve problems. Rather than rely on random mutation, as with traditional plant breeding, gene editing will precisely target "disease susceptibility genes, making new varieties disease-free,"

---

214   Communication from Dr. Art Schaafsma (January 2017).

he says. Dr. Folta is investigating a solution for citrus greening, a bacterial disease that is threatening citrus crops around the world.[215]

How quickly will the public, government regulations, and the limits of our imaginations allow science to follow this new frontier? Where concerns over pesticide use and the environment are verified by peer reviewed science, farmers too want resolutions. I'm confident our OSCIA members will be clamouring to continue farm trials to learn of new advancements to solve our Ontario soil and crop challenges.

Farm Shows Are Where It's At

OSCIA strives to be highly visible at events attended by farmers, such as the International Plowing Match and Rural Expo (known as the IPM) and Canada's Outdoor Farm Show (COFS). The IPM has been held since 1913 and moves from county to county each year. These venues are great opportunities for OSCIA outreach—engaging conversations with farmers and industry representative about applied research and program details.

The International Plowing Match

Each year throughout the 1990s, I loaded up my vehicle with the paraphernalia needed to set up the OSCIA exhibit space under one of the tents at the IPM. Since I owned a four-wheel drive Ford Explorer with considerable cargo space, it was logical that I would haul the supplies and banners. Furthermore, with its 4X4 capability, the Explorer could navigate through the soggy fields and softer ground that often follows September rains—or so I thought, until I got to the Haldimand County location of the IPM in 1996. Wow! Torrential rains the previous week left roads at the site deeply rutted and impassable to even 4X4s. Tractors were required to tow pickup trucks. I recall being towed by an International 2+2 (a large four-wheel drive tractor) to the tent where OSCIA had acquired space. Somewhere, someone has a video of our entourage schlepping from rut to rut to

---

215   Folta, K. "Get ready for gene editing." Citrus Industry (October 2016). www.crec.ifas.ufl.edu/extension/trade_journals/2016/2016_October_gene.pdf (Accessed Aug, 31, 2017).

our final destination. There was a permanent kink in my Ford frame after that IPM. Fortunately, the sun shone down on many other IPMs and it was an absolute treat to be able to attend these week-long festivities. In recent years, OSCIA is represented by the local county/district associations at the IPM.

> Another appealing factor was that the show would focus strictly on the business of farming

Canada's Outdoor Farm Show

> *"At the 1993 summer meeting, hosted by 1994 OSCIA President Victor Roland, a meeting was held in their farmhouse kitchen where a motion was passed to support the endeavour of Ginty Jocius to start Canada's Outdoor Farm Show. Vic and I were recognized by Ginty for our volunteer support during those start-up years."*
>
> -Allan Brown, 1995 OSCIA President (Simcoe North County)

In 1993, OSCIA was approached by Guelph entrepreneur Ginty Jocius, who proposed creating a permanent farm show site where field trials could be conducted, lending to continuity of field research results. The first discussion occurred during the summer meeting at Victor Roland's farm. Executive members were asked to sign a confidentiality agreement, likely the first time an OSCIA executive was exposed to such a legal instrument. But when working with private entrepreneurs, it is essential to keep new business ideas under wraps to ensure the competition doesn't get an edge.

Doug Wagner, secretary-manager of OSCIA at the time and now the president of Canada's Outdoor Farm Show, says that the executive took some heat from the board members and local associations concerned about competing with the IPM. However, OSCIA was to be brought on board as a founding partner, providing our association with a grant and permanent site for crop trials. Another appealing factor was that the show would focus strictly on the business of farming. The public announcement was made in September 1993 and the inaugural exhibition opened a year later near Burford, Ont.

The OSCIA planning committee continues to be involved in planning each year's show activities and designing the layout at the OSCIA designated exhibit area. Since COFS is now located near Woodstock in Oxford County, it has been quite convenient for the local association's Cathy Dibble to take a lead role in co-ordinating OSCIA's exhibit area.

The partnership with COFS has been a truly collaborative effort that also benefits from the efforts of OMAFRA crop technology staff at the crop demonstration site. A permanent site has allowed the planting of perennials including trees and shrubs that are contained in the conservation forestry section of the southeast corner of the site. A systematic crop rotation with cover crops is also a great showcase. With recent breakfast sponsors such as Sylvite Agri-Services, OSCIA members can start their day off at the show with a free breakfast and good conversation with familiar faces from OSCIA and OMAFRA. It is becoming quite common to see farmers lined up to converse with one of their favourite advisers about a crop challenge or new soil management technique. The COFS slogan "Where Crop Farmers Meet" is fitting. With expanded tent space, new expert stations, and the catchy tagline "Grass Roots and Dirty Boots," the 2016 event was a banner year for visitors to the OSCIA/OMAFRA showcase. Kudos to the planning committee for their creative thinking.

Working in Concert

OSCIA is a member of numerous organizations to strengthen our efforts as an industry and collaborate with communication efforts. Skills and focused expertise provide additional benefits to serve the needs of the agricultural industry. These organizations include:

1. Farm and Food Care Ontario

Farm and Food Care Ontario's mission is to "proactively work with Ontario farmers and food partners to ensure public trust and confidence in our food system.

"Farm and Food Care Ontario provides credible information on food and farming to non-farmers, and engages the agriculture community through training, projects, and resource development in three key areas:

- Farm animal care
- Farming and the environment

- Food and farming resources."[216]

2. AgScape

AgScape creates "agriculture and food literacy and career engagement by supporting student learning." This is achieved by:

- "Creative programming which connects food, farming and health, expanding the profile of Ontario's agri-food sector;

- Collaborating with partners to deliver education programming that excites students and teachers, and encompasses all food production systems in Ontario; and

- Connecting classrooms with agri-food perspectives that increase understanding, encourage critical thinking, and stimulate dialogue."[217]

3. Ontario Agri-Food Technology (OAFT)

OAFT's mission is to "provide leadership and co-ordination in utilizing technology to generate wealth and sustainability for the agricultural and food industries of Ontario by focusing on four key action areas:

- "Encourage and support collaboration among public and private research groups to enhance the development and acceptance of products we produce;

- Development, support, and co-ordination of research programs amongst Ontario's institutions; thereby providing new product opportunities, ensuring innovation, efficiency, and focus, and establishing new research partnerships and business opportunities, many of which will be in non-food areas;

- Creating a skills and expertise database of research scientists in biotechnology and bioproducts to help ensure that the best scientific expertise is used on a specific project. Maintaining this resource

---

216 Farm and Food Care Ontario, http://www.farmfoodcareon.org/about-us/ (Accessed January 17, 2018).

217 AgScape. About. https://agscape.ca/about (Accessed January 17, 2018).

will also facilitate communication regarding potential research projects and focus research on relevant applications; and

- Securing funding to support research and development of new technologies and importantly, helping to find financing for companies to produce their products."[218]

3. Ontario Forage Council

"The mission of the Ontario Forage Council is to provide leadership to the Ontario forage industry through communication, research, technology transfer, market development, advocacy, policy development. and bringing stakeholders together."

The council's objectives are:

- "To act as the primary forage information source;

- To collaborate with other Ontario agricultural organizations for the development of information and educational materials as well as other service related projects for forages;

- To assist in the objective development of research priorities, co-ordination of research projects, and patent registration related to forage production, management, and marketing;

- To promote the positive environmental impacts of forage and grassland on soil and water conservation, carbon sequestration, reclamation, wildlife habitat, and aesthetic value;

- To identify, receive, and distribute funds to support development of the Ontario forage industry; and

- To develop and promote market innovations at the local and export levels."[219]

---

218 OAFT. About Us. http://oaft.org/about-us/ (Accessed January 17, 2018).

219 Ontario Forage Council. Our Role. http://www.ontarioforagecouncil.com/home/our-role (Accessed January 17, 2018)

4. Canadian Forage and Grassland Association

"The main role of the CFGA is to uphold the robust hay and forage industry and realize the potential of the domestic and export forage market."

- "Our excellent weather conditions and abundance of fresh water, clean air, and fertile soil enable the production of the highest quality hay in the world. Several types of forage are produced in Canada, including bromes, fescues, timothy, alfalfa, clover, and orchard grass.
- Our production supports the dairy, beef, sheep, and horse industry domestically and abroad.
- CFGA is the umbrella organization for the provincial and regional councils, providing a co-ordinated voice for the sector."

The board of directors is focused on:

- "Raising the profile of the industry with its work on innovation;
- Marketing and sales;
- Increasing profits for member producers; and
- Working on behalf of forage exporters, helping navigate issues such as transportation costs, currency rates, protocols, energy costs, and market demands."[220]

5. Soil Conservation Council of Canada (SCCC)

"Soil produces 95 per cent of the food we eat and is the foundation on which our civilization rests. We must continually strive to improve practices, or our foundation will crumble and along with it, humanity."

- Alan Kruszel, Chair, Soil Conservation Council of Canada, and 2015 OSCIA President (Stormont county)

---

[220] Canadian Forage and Grassland Association, http://www.canadianfga.ca/ (Accessed January 17, 2018).

The Soil Conservation Council of Canada's goals are to:

- "Provide leadership in the protection and care of Canada's soil and related resources;

- Improve the level of understanding and awareness of the importance of soil and soil health among all Canadians and to increase their support of soil conservation;

- Facilitate communications among various stakeholders for work relating to soil and related resources;

- Communicate to the general public those policies, programs, or activities that affect the sustainable use of Canada's resources;

- Encourage the development of policies, production methods, and management systems for agricultural land use that enables sustainable use of soil and related resources; and

- Work collaboratively with stakeholders in the delivery of soil related conservation programs."[221]

6. Canadian Roundtable for Sustainable Crops (CRSC)

The Canadian Roundtable for Sustainable Crops supports "Canada's grains sector in adapting to and thriving in an evolving operating environment. The CRSC works to:

- "Ensure the agriculture industry develops a unified, proactive approach to grains sustainability in Canada;

- Explore and create a national, cross-commodity sustainability solution that is accepted by producers and has credibility with key stakeholders;

- Research and develop farm-level sustainability indicators that are relevant to Canadian production practices (environmental, social, and economic);

---

221   Soil Conservation Council of Canada. About Us, http://www.soilcc.ca/about-us.htm, (Accessed January 17, 2018).

- Inform and engage Canadian grain producers on the topic of grains sustainability; and

- Represent Canada and improve its reputation in crops sustainability by communicating the sector's 'good news' story."[222]

Will we be Amassing Biomass En Masse?

We've recognized how soil is healed and conserved with the use of hay, pasture, and the growing of perennial crops such as alfalfa. But if a farmer has no livestock or market outlet for bales of hay, what can one do? Certainly, cover crops/green manure can help. Fortunately, the twenty-first century is opening doors to new opportunities.

Leading up to 2010, there was burgeoning interest in growing perennial plants as energy crops. When the political announcement was made to phase out all coal-fired electrical plants by 2014, the agricultural industry began to investigate the potential opportunities for purpose-grown crops for energy, as there were many questions to be answered and feasibility options to consider. The Ontario Federation of Agriculture (OFA) took the lead in crafting a proposal and on June 29, 2010, the federal government announced a $2.4-million contribution to OFA. The funds came through the Canadian Agricultural Adaptation Program (CAAP), delivered in Ontario by the Agricultural Adaptation Council (AAC). OSCIA would assist with on-farm research and development with an agreement with OFA signed September 2010.

The rationale for feasibility studies was solid. A work plan was developed entitled, "A Transformative Project to Generate Energy for Ontario by Developing an Innovative Agricultural Biomass Value Chain." In the OFA/OSCIA agreement, the summary of our quickly evolving investigation focused on three components:

---

[222] Canadian Roundtable for Sustainable Crops, http://sustainablecrops.ca/ (Accessed January 18, 2018).

- **"Seizing Opportunities**: Under the Ontario Environmental Protection Act, the Ontario Cabinet enacted Regulation 496/07, which states: "The owner and the operator of each of the following generating stations shall ensure that coal is not used to generate electricity at the generating station after December 31, 2014: 1. Atikokan Generating Station, located in the Township of Atikokan, 2. Lambton Generating Station, in the Township of St. Clair, 3. Nanticoke Generating Station, located in Haldimand County, 4. Thunder Bay Generating Station located in the City of Thunder Bay.' This directive is creating a large market for alternative fuel sources, as these generating stations, owned and operated by Ontario Power Generation (OPG), are expected to continue to be needed to supply Ontario's electrical needs during peak periods. OPG expects that the use of existing assets with new fuel will be economically viable in comparison to other alternatives. Collectively, OPG estimates they represent a market for 2,000,000 metric tonnes of biomass pellets by the time coal is phased out in 2014. There are essentially two sources for this biomass material: (a) woody biomass from the forest products industry, and (b) biomass from agriculture (purpose-grown and residual). For the Atikokan and Thunder Bay Generating Stations, given their location in the northwestern part of Ontario, OPG favours biomass (a); for the Lambton and Nanticoke Generating Stations, given their location close to agricultural land, OPG favours biomass (b), provided that it is available and economically viable. This decision to phase out coal and seek an alternative fuel has created an immense opportunity for the agricultural sector in Ontario to respond and produce the material needed. On the basis of a yield of four tonnes per acre, the amount of agricultural land required could approach 250,000 acres and have a farm gate value of $150,000,000. After the farm gate, further value will be added at the aggregator stage in forming the dense pellets required for combustion. This opportunity could be truly transformative for Ontario agriculture, leading to the emergence of an entirely new family of crops, a new stable market, and a new industry involved in the harvesting and storing of purpose-grown biomass and the aggregation of combustible pellets. In addition to the primary market represented by OPG, the cement industry in Ontario is interested in purpose-grown biomass as an alternative to the coal now in use in firing the kilns used for making cement. Plus, the greenhouse sector in Ontario is interested in purpose-grown biomass as a fuel source for the boilers needed to heat their greenhouse structures. A number of Ontario greenhouses are now burning woody biomass which shows the adaptability of the boiler technology to alternative fuels. In order to seize this opportunity, many questions need to be investigated and answered as elaborated in this application. The approach of this project will ensure those questions are answered in a co-ordinated and methodical way and that the results are made available to growers across Ontario so that they can make informed decisions. Because biomass has a relatively low value/bulk ratio, it is costly to ship across significant distances. By its very nature, therefore, the emergence of a purpose-grown agriculturally-based biomass sector must occur regionally, close to the facilities where it is proposed to be burned. Therefore, this is an opportunity first and foremost for farmers in Ontario.

- **"Pathfinding Solutions to Adapt and Remain Competitive**: Pathfinding means looking at different options to prepare the sector to face the future and remain competitive. Because purpose-grown biomass in Ontario is a potentially new commercial crop, there are many questions to be answered in determining the optimal conditions for growing, harvesting, storing, and pelletizing the crop economically. These questions are elaborated in the application. This project is intended to provide as many answers as possible by looking at different options for the future related to the choice of species (switchgrass, miscanthus, big bluestem, etc.); the choice of soil conditions; the optimal times to plant and harvest; the extent to which nitrogen application is needed; pests and diseases to which stands may be susceptible and how to control them; ways to store the harvested crop; extracting nutrients that are undesirable for combustion and returning them to the soil;

how to aggregate at least cost; what type of dies to use in pelletizing; and so on. In addition to the technical aspects of growing, storing and aggregating agricultural biomass, there are also organizational issues associated with the emergence of a new sector. One of the pathfinding initiatives relates specifically to options for organizing the aggregators' role in the value chain. One option that will be investigated is the possibility of grower ownership on a co-operative basis. In addition, given the cost of customized seeding and harvesting equipment, the industry is expected to be a boon for custom operators who provide those essential services to growers on a fee-for-service basis to make best use of their capital investment in the equipment assets.

- **"Piloting to Find Solutions to New and Ongoing Issues to Adapt and Remain Competitive**: Piloting means the testing of ideas or approaches to see if they are effective enough to use in everyday applications in the sector. Closely tied to pathfinding, the greatest proportion of project spending relates to field and aggregator pilots to obtain hands-on experience with purpose-grown biomass in parts of Ontario with different soil and climate characteristics. This investment is important to getting more Ontario farmers experienced with the crop and able to provide test sites to which other farmers can go to examine the crops and learn about their agronomic traits. It will also help identify success factors that contribute to the best stands, the best yields, and the best energy values with the least establishment cost and lead-time to be in production. The experimental design of the aggregator trials for pelletizing will also be designed as a combination of pathfinding and piloting, testing different approaches to pelletizing with different types of dies, different moisture levels in incoming material, different methods for destining and preparing the incoming material."[223]

---

223 "A Transformative Project to Generate Energy for Ontario by Developing an Innovative Agricultural Biomass Value Chain." OSCIA/OFA agreement, signed September, 2010.

The work plan directed our efforts in the following areas:

- Literature reviews, including global analysis;
- Economic assessment, optimization, testing, and scale-up;
- Life-cycle analysis;
- Aggregator structure including a co-operative arrangement;
- On-farm research and development on up to nine hundred acres geographically dispersed across Ontario; and
- Communications.[224]

> Agreements were finalized with twenty-eight producers covering nearly 725 acres

Pathfinding and pilot testing for on-farm research is where OSCIA jumped in to help. With close to $1 million made available as part of the agreement to support leading-edge field trials, OSCIA put out the call for expressions of interest. Agreements were finalized with twenty-eight producers covering nearly 725 acres. Provision of cost-share to cover the expense of establishing purpose-grown crops was fully justified given the uncertainty of markets and the fact that it took two to four years to establish optimal yields from these perennial crops.

Research at the University of Guelph led by Dr. Bill Deen examined the issues of growing "energy crops" at the participating farms as well as the university's **own** sites. According to the OSCIA's final project report, many producers struggled to develop solutions to the agronomic and productive capacity challenges involved with purpose-grown biomass crops. This can likely be attributed to the slow development of markets and producers lacking the time or interest in developing crop experience where the financial payoffs are unknown. While many basic agronomic questions were addressed and biomass productive capacity was developed, several key issues remain. They include: questions of variety selection and suitability for different regions;

---

224   Ibid

effective grass and weed control, and fertility management for optimal economic returns.

Despite the challenges, the final report on OSCIA's portion of the project recommended that the work should continue.[225]

As the research was underway, a damper was imposed on large-scale production of biomass crops for energy. The price of natural gas plummeted, and the economics of purpose-grown crops on farms for combustion has been limited to a few greenhouse operations or other local facilities where natural gas is not available. OSCIA collaborated with OMAFRA's Best Management Practices (BMP) Verification Program to investigate efficiencies and combustion emissions of small-scale combustion units. Headed by OMAFRA engineer Terrance Sauve, the project evaluated numerous types of biomass feedstock that was burned as fuel in a variety of hydronic (radiant) heaters installed on a greenhouse farm north of Kingston, Ont. Key findings from the study suggest that the highest efficiency and lowest emissions are achieved when the boiler is equipped with oxygen sensors that can automatically adjust the air-fuel mixture and where the surplus heat is captured in reserve tanks for future use.[226]

Perhaps biomass for energy production is still slightly ahead of its time? When the next spike in energy costs arrives, we will be much better prepared to evaluate choices, economic options, and implement the best soil fixing remedies.

Throughout the biomass for energy investigation, another interesting innovation emerged. OSCIA was involved in organizing a national biomass conference in Ottawa in August 2013. One of my take-home messages from

---

225 Engbers, H., Deen, B. "Field Scale Agricultural Biomass Research and Development Project Final Report (November 2013) https://www.ontariosoilcrop.org/wp-content/uploads/2015/11/biomass_final_report-january2014-h.engbers_b.deen.pdf. (Accessed February 5, 2018).

226 Sauvé, T. "Agricultural Biomass Heating." www.ontariobiomass.com. http://www.ontariobiomass.com/Resources/Documents/Ag%20Biomass%20Day%202015/Ag%20Biomass%20Day%202015%20Presentations/13%20-%20Agricultural%20Biomass%20Heating%20-%20Terrence%20Sauve,%20Ag%20Biomass%20Day%202015.pdf (Accessed Feb. 5, 2018).

that event was that Ontario farmers stand to benefit from new markets opening up to serve many industries that are seeking alternatives to fossil-fuel-based chemicals used in their manufacturing processes. And I'm not just talking about ethanol, which is made primarily from corn and has been used for years as a fuel additive. The next big opportunity on the horizon may come from producing industrial sugars from cellulose crops—extracting value from plant residue, especially stover, the leaves and stalks of corn, soybeans, or cereal crops left in the fields after harvest.

In Ontario, the Cellulosic Sugar Producers Co-operative (CSPC) was established to help the industry develop new markets for crop residue materials like corn stover. The CSPC and its partners are working to develop new high-value biochemical and biofuel applications for cellulosic sugars.[227]

Of course, science is also keeping a watchful eye on what quantities of crop residue can be removed while maintaining a satisfactory threshold of organic matter for sustainable soil health. There are a multitude of these scientific investigations underway in many jurisdictions around the world. In Ontario, the Ontario Federation of Agriculture (OFA) has been leading the way, assisting in several feasibility studies associated with harvesting crop residue and biomass production.[228]

There are unlimited opportunities to support innovation with research and development in agriculture. This chapter has provided a brief glimpse of how many partners work together, each with a unique and significant role to advance the goalposts for sustainable agriculture. The next chapter provides an in-depth look at sustainable agriculture where many partners are collaborating to piece together the complex requirements for further transparency through verification processes.

---

227   Cellulosic Sugar Producers Co-operative, https://www.cspcoop.com/ (Accessed February 5, 2018).

228   OFA website. Bioeconomy Overview. https://ofa.on.ca/issues/bioeconomy/

Chapter 13

THE REAL MEANING OF SUSTAINABLE FARMS AND FOOD

"Finding balance between the economics, agronomics, and conservation within Ontario agriculture has always been a challenge. Throughout its many years, OSCIA has maintained a practical approach to achieving this balance by transferring knowledge to growers through both research/extension to our grower network, but just as importantly, a peer-to-peer network."

- Greg Kitching, 2002 OSCIA President (Halton Region)

What word, over the past thirty years, has been the most misused, misunderstood, abused, and exploited? For me, the word is "sustainable." I have limited references to it until this point because it is important to establish context for how, why, when, and where the word should be used.

The definition of sustainable development that has withstood the test of time goes back to a high-profile report called *Our Common Future* issued in 1987 by the World Commission on Environment and Development of the United Nations. Named after the commission chair, Ms. Gro Brundtland, former Prime Minister of Norway, the Brundtland Report identifies three pillars of sustainability: environmental protection, social equality, and economic growth.[229]

---

229   "Our Common Future: Report to the World Commission on Environment and Development." Chairman's Foreword. www.un-documents.net/our-common-future.pdf (Accessed May 25, 2017).

The United Nations (UN) continues its leadership today with reference to seventeen Sustainable Development Goals (SDG) to be implemented around the world by 2030. Canada has embraced the UN's seventeen SDGs, which delve much deeper into details on the original environmental, economic, and social components of Brundtland report.[230] Agriculture and food are integral to sustainable development, opening new opportunities for co-operation and collaboration within Canada.

To provide a vision for how Ontario farmers and food manufacturers should respond to growing demands for sustainable farm production, Ontario farm and industry leaders brought forward their vision in 2015 with a document called *Farm, Food and Beyond - Our Commitment to Sustainability*. Building upon twenty-five years of environmental leadership by Ontario farm organizations, the document broadened the industry's commitment to tackling global challenges based on the following understanding of sustainability:

"Sustainable development means a strong focus on environmental quality. It means ensuring that current human activities—including agriculture—do not cause irreparable damage to the **natural environment** for future generations. It means maintaining and improving the quality of water, air and agricultural soils. It also means using agriculture, where possible, as a mechanism for environmental improvement—supplying renewable resources, for example, and converting atmospheric carbon dioxide into soil organic matter. Sustainable development is recognized as having two other dimensions: **social** and **economic**. For agriculture, social sustainability includes an adequate supply of safe, nutritious food at affordable prices—for both Canadians and the rest of the world's 7.3-billion citizens. The Brundtland Commission's emphasis on needs of the poor is especially significant. Environmental improvement must be balanced with the daily food needs of the billions of impoverished people across the globe. Sustainable development also means a decent quality of life and safe working/living conditions for those who work in agriculture—farmers, their families, employees—and others in the agri-food system. Social sustainability includes a commitment to respectful and responsible treatment of farm animals, which form a critical component of

---

230   The 2030 Agenda for Sustainable Development, http://international.gc.ca/world-monde/issues_development-enjeux_developpement/priorities-priorites/agenda-programme.aspx?lang=eng (Accessed May 10, 2018)

the Canadian and world food system. Farm animals are important in providing a high-quality, nutritious diet for many people. They play essential roles in converting many plant species that are not directly digestible by humans into food products (e.g., meat, milk, eggs), and that are especially valuable in maintaining/building agricultural soil quality (e.g., alfalfa and other forage crops). Food supply, quality and safety, and animal care are core elements of agricultural sustainability. The economic component is equally critical. Agriculture and food-supply chains directly dependent on agriculture cannot be sustainable if there is not a reasonable financial reward from farming that enables an appropriate standard of living for farm families. Canadian farm families should enjoy the same quality of life and financial well-being as other Canadian families."[231]

> With the many different components of sustainability, best practices related to soil health and soil care are essential and are the foundation of food production

If one examines the multitude of programs which OSCIA has delivered over the past thirty years, could one not conclude that they each contained an essential component of sustainability? In fact, most participants would suggest that the EFP does an excellent job of drilling down into the technical details of environmental sustainability. Does it not define Best Management Practices (BMPs) and make each producer-participant aware of how their farm can be environmentally sustainable? In the broadest sense, soil fixers have been an integral part of sustainability in their efforts to eliminate soil erosion, improve soil quality, and reduce contamination of watercourses. Being stewards of the natural resources in and around our farms are also part of it. With the many different components of sustainability, best practices related to soil health and soil care are essential and are the foundation of food production. Our civilization depends on it. We do, however, require a more transparent method of verifying and reporting the

---

[231] "Farm, Food and Beyond - Our Commitment to Sustainability." Sustainable Farm and Food Initiative, https://static1.squarespace.com/static/5a4fc47f1f318d07aef77163/t/5a53a4f2652dea1af95a13e7/1515431161491/OurFarmSustainableAgenda-LR.pdf (Accessed May 17, 2017).

conditions of our soil. A comprehensive sustainability plan that is accepted and recognized around the world will position Canada as a world leader.

Growing Your Farm Profits

What about the social and economic pillars of sustainability? Farming in a socially responsible manner, making a profit, investing in upgrades to equipment and buildings, supporting the family and community, and developing a succession plan are as important to sustainability as environmental responsibility. The requirements of social and economic sustainability can be integrated into a farm's comprehensive business plan. But convincing farmers to develop detailed written business plans for their farms had attracted little participation in the past. There were perhaps several thousand business plans formally drafted by leading farmers up to 2006. Farm leaders and government experts were concerned about the low numbers of farm business plans and began to explore the barriers and challenges.

In the fall of 2006, the Agricultural Management Institute (AMI), led by Alliston, Ont., potato grower Dr. Peter VanderZaag, assembled an advisory panel that held five think tank workshops across Ontario and consulted with more than seventy-five key industry influencers about farm business management and succession planning. The process continued with the Farm Management 2020 conference that included over 120 key industry stakeholders who discussed solutions to the business planning issues identified by workshop participants. Participants came from throughout the agricultural sector including farmers, supply businesses, agricultural education, financial lending, research, government, and more. The diverse group included people from a variety of backgrounds, from young adults in their twenties to industry veterans in their seventies. The group called for development of a program that made business planning understandable and that was practical for them to use. The goal was to create a farm business management program akin to the Environmental Farm Plan program (EFP), which takes participants through a series of self-assessment worksheets and helps them develop action plans to address challenges. The group noted that the EFP, which had attracted more than ten thousand participants in the previous two years alone, had been successful in getting farmers to voluntarily take action on environmental issues.

A key recommendation of this consultation was that AMI undertake a process to validate, pilot and test business planning templates and tools for farmers. The AMI advisory panel approached the OSCIA to work with OMAFRA to develop a template similar to the EFP model. The intent of the program was to increase the level of business planning among farmers in Ontario, and to help increase profitability in the sector. It aimed to help farmers manage their business risk by arming them with the tools needed to plan for success and anticipate opportunities, react to changing circumstances and avoid failure.[232]

The Growing Your Farm Profits (GYFP) program was the result. Topic areas for GYFP included:

- Marketing
- Production
- Financial Management
- Human Resources
- Social Responsibility
- Succession Planning
- Business Structure
- Business Strategy

The GYFP included the following features of the EFP:

- Self-directed risk assessment
- Guided action plan development
- All commodities
- All farm systems or philosophies
- Confidentiality

The proposal calling for OSCIA to deliver a program that was not about growing crops or fixing soil did cause some angst among our board members

---

232 OSCIA Archives.

back in 2006. What did business planning have to do with soil and crop management? Diversification into the broader scope of farm business planning was not taken lightly by the board. Was this just another program for OSCIA to acquire funding?

At the summer board meeting in Peterborough, Ont., an idea blossomed suggesting that business and economics could fit within OSCIA's mandate. After all, why would our progressive and conservation-minded farmers continue to invest in fixing soil and protecting waterways if they didn't have a sound financial plan that also included a succession plan for future generations?

Discussions continued through the end of 2006 and into early 2007. There was consensus that, if handled properly, adding a business-management focus to OSCIA's mandate would be a positive development. In February 2007, the OSCIA board approved a motion to pursue funding for a pilot project.[233]

The pilot was to test multiple facets of the GYFP program content, delivery, and design. The workbook and workshop were developed by OSCIA and OMAFRA with funding through the AMI. The program process entailed four steps:

Step 1: Taking stock: self-assessment

Step 2: Setting goals

Step 3: Setting priorities identified in the self-assessment that support your goals

Step 4: Creating actions that achieve the priorities, using available resources

The pilot program engaged up to 140 producers in seven locations that each represented a unique geographic area of Ontario with unique set of agricultural characteristics. The pilot was facilitated by EFP workshop leaders in the following counties/districts: Lambton; Bruce; Middlesex; Niagara; Northumberland; Prescott and Russell; and Nipissing.

---

233   OSCIA Director Minutes. Feb. 5, 2007. OSCIA Archives.

There were two styles of workshop to test. One was a traditional style that included introductory presentations by the facilitator for each topic, followed by each participant completing the assessment and action plan. The second style included experiential learning that engaged participants in simulation exercises in business decision-making to solve challenges that may arise on a typical family farm. Follow-up surveys indicated the experiential learning style was more effective.

Participants were pleased with the workshop and encouraged the leaders to ensure the workshops would be delivered on an ongoing basis. Some of the comments from the feedback sheets are noted below:

- "The books were very well laid-out and allowed for seamless transfer of information. Having the ability to rate the business and related activities was a good process to go through.
- Very good two days!
- Good program to get us thinking about our businesses.
- Keep program going as it has merit.
- Right amount of time for self-examination. Good idea. Good exchange of thoughts and ideas among group.
- The staff was excellent. Good second day session. Good flow. Entire group participation. Very well done. Resources provided are good.
- Great presentation of examples - how to deal with day to day challenges.
- Excellent!!
- Great workshop that can be applied to business and life in general.
- One of the best workshops I've done!
- It made me think and it made me laugh and it was fun!"[234]

---

234   OSCIA archives

Recommendations to Implement GYFP

The following recommendations were made based on the results of the pilot workshops:

- Proceed with a full experiential program.
- Train three to four regional facilitators/business management coaches to work with local OSCIA representatives who would act as assistants.
- Encourage all members of the farm management team to attend the workshops.[235]

By all accounts, the pilot project for GYFP was a smashing success and set the stage for implementation of a full provincial program. At the OSCIA 2008 summer meeting in Sault St. Marie, Ont., the board of directors supported a motion "that OSCIA proceed with plans to initiate a Growing Your Farm Profits program under the Growing Forward program with OMAFRA."[236] Our hosts for that meeting, Norma and Murray Cochrane (2009 OSCIA President), rose to the occasion in typical OSCIA fashion and organized a hallmark event near Thessalon, Ont., that marked the launch of an exciting new program to emphasize the importance of business planning for Ontario farmers.

Expansion of GYFP

A separate team within OSCIA was set up to administer and facilitate GYFP. Workshop leaders were recruited, some from the pool of experienced EFP personnel in each area. At first, some participants were a bit uncomfortable with the experiential style of workshop that required them to interact and communicate

---

235 Fitzgerald, S. Fitzgerald & Co. "Farm Business Planning Self-Assessment and Action Planning Workshops, CanAdvance Project (ADV #470 Pillar 1: Industry Led Solutions to Emerging Issues, Final Report, August 29, 2008) OSCIA Archives.

236 OSCIA Directors Minutes, Aug. 18, 2008. OSCIA Archives.

with the other workshop participants. But once the ice-breaker exercise would get workshop participants talking and laughing, they were quickly put at ease.

GYFP got into full swing across the province under Growing Forward, a federal/provincial/territorial initiative that ran from 2008 to 2013. There were 4,183 producers who took GYFP in Growing Forward. The next five-year agreement, Growing Forward 2, engaged close to an additional 3,000 participants.

Compared to the environmental projects of the EFP, the dynamics are different for farm business plans that involve proactive thinking. Changes are influenced by family dynamics and legal structures that often involve professional farm business advisers, accountants, lawyers, and bankers. Multiple interests must be considered. Good communication and a thorough analysis of finances and business processes are paramount.

Financial assistance has been made available for some of the professional fees but under Growing Forward 2 (GF2), business planning is but one of the focus areas considered in the merit-based approval process that includes:

- Environment and climate change adaptation;
- Assurance systems (food safety, traceability, animal welfare);
- Market development;
- Animal and plant health;
- Labour productivity enhancement; and
- Business and leadership development.[237]

Does the above list not appear to cover off the essential components of sustainable farming? The economic and social components of the original definition of sustainability seemed to be staring us in the face.

---

237   Growing Forward 2 Quick Reference Guide. www.ontariosoilcrop.org/wp-content/uploads/2017/02/GF2-QRS-FundingLevels.pdf (Accessed Sept. 6, 2017).

The Buzz Continues on Sustainability

Interestingly, throughout all the discussions among farm leaders, government partners, and OSCIA board members involved in the development of GYFP, there was initially no reference to the word "sustainability." We had not clued in to how GYFP could complete the sustainability package by blending the economic and social components of farm business planning with the environmental complement of EFP. Nor was anyone asking.

Even our Growing Forward 2 agreement did not reference sustainability. The topic of sustainability among food manufacturers and retailers didn't come across my desk until well into 2012, when talk about sustainable sourcing of food began percolating into the general news media. Buyers around the world were becoming interested in corporate social responsibility across the entire value chain. Shareholders were demanding it and no business wanted to be seen to be violating emerging standards of environmental protection, labour conditions, animal welfare or food safety.

What did that mean? Well, it depended on the commodity and country. For example, the International Trade Centre Standards Map provides "information on over 210 standards, codes of conduct and audit protocols addressing sustainability hotspots in global supply chains."[238]

Issues of labour standards are a high priority in the developing world where commodities such as coffee or chocolate are major exports. It soon became apparent that there were a multitude of sustainability schemes out there, each with its own set of strengths and limitations and some more credible than others. Several hundred standards, definitions, and criteria are a recipe for confusion and label fatigue along the value chain, never mind the perplexity it would cause for consumers. What are their true motives? Was one company simply attempting to obtain a market advantage by claiming their product to be more sustainable than their competitor's? What kind of verification system did they have?

I began to quiz other colleagues about their understanding of the evolving requirements for sustainability. I was convinced that EFP had the

---

238   Welcome to Standards Map. International Trade Centre. www.standardsmap.org/identify (Accessed May 30, 2017)

environmental best practices covered in spades, but we did not have a transparent verification system, nor a mechanism for certification, if required. I was determined that Ontario's agricultural industry, with all its achievements, and successes, did not have to reinvent environmental sustainability.

OSCIA, under advisement from other industry leaders, engaged the George Morris Centre to undertake an investigation to test the Ontario EFP against the requirements of the food manufacturing and retail sectors. Dr. Al Mussell and Dr. Claudia Schmidt headed up the work that included a literature search and interviews with food processors, manufacturers, retailers, and other key industry leaders. Their report concluded:

"The EFP has achieved significant success in creating education and awareness of agricultural environmental issues, and material success in on-farm environmental improvements and remediation. At the same time, producers are facing pressure to demonstrate the sustainability of their on-farm practices. With a well-established EFP and extensive participation, Ontario should be well positioned to address these demands. The results of this study suggest a potential starting point to meet demands by using EFPs that are already in place; most jurisdictions outside of Canada lack a similar platform. There is an important opportunity for further work to leverage this and assess the feasibility for the EFP as an important sustainability tool that can improve Ontario's and Canada's positioning as a key supplier of food that meets increasing sustainability expectations."[239]

Industry leaders in Ontario continued to discuss the challenges of reducing redundancy, confusion and the growing paper burden of regulation. Organizations taking the lead along with OSCIA were the Ontario Federation of Agriculture, Christian Farmers Federation, Farm and Food Care, Ontario Agri-Food Technologies, Presidents' Council, National Farmers Union, University of Guelph, and Provision Coalition with its members from food and beverage manufacturers.

---

[239] Schmidt, C. et al. "Potential Role of the Ontario Environmental Farm Plan in Responding to Sustainability Demands of the Agri-food Supply Chain." p. 67. George Morris Centre. (2013) www.georgemorris.org/publications/OSCIA_-_EFP_-_Final_Report_August_29.pdf (Accessed May 30, 2017).

These organizations formed the initial steering committee to tackle the challenges of providing food and fiber in a sustainable manner. Government partners from OMAFRA and Agriculture and Agri-Food Canada (AAFC) were brought on board to participate in the working group. The result of their efforts is outlined in *Farm Food and Beyond - Our Commitment to Sustainability*. Here are the key points:

- "Ontario agriculture must increase its production of food ingredients per unit of farmland by at least three per cent per year.

- Environmental Farm Plans will be transformed into Sustainable Farm and Food Plans to encompass the other two main pillars of sustainable development—economic and social sustainability.

- Social sustainability includes a commitment to the respectful and responsible treatment of farm animals.

- Sustainable Farm and Food Plans will include a special focus on climate change.

- Sustainable Farm and Food Plans will recognize all opportunities to reduce wastage.

- Efforts to enhance farm sustainability will be developed in close co-operation with Ontario and Canadian food industry partners, the largest users of Ontario grown farm products.

- Ontario-based efforts to produce new, renewable, bio-based non-food products as replacements for traditional fossil fuel-derived materials will continue to be emphasized.

- Sustainable Farm & Food Plans will be developed with three objectives:
    - To guide farmers in the identification of needs/opportunities for improvement in sustainability;
    - To assure and inform the general public about these transformations; and

- ○ Help address growing requirements by food manufacturers and retailers for assurance that farm products have been produced in sustainable ways.

- • Sustainable development depends on the recognition of traditional Ontario farmer knowledge and acquired knowledge, and the acceptance of new, advanced technologies.

- • Sustainable Farm and Food Plans will be whole-farm focused, instead of commodity-specific. These will be based on solid science and reflect a 'systems approach,' which is fundamental to good farm management.

- • Our commitment for the next twenty-five years will include continuously improving communication with other Ontarians/Canadians."[240]

> Could there be a common platform that would integrate each farmer's efforts, minimizing overlap and duplication?

The Sustainable Farm and Food Initiative (SFFI) steering committee and working group, chaired by Dr. Gord Surgeoner, has been meeting with stakeholders since 2015 to work toward uniting the industry and capitalize on the achievements already underway in Ontario. The work is building upon the investment by the industry in EFP, GYFP, and other related programs to work on a common platform to log and retrieve assessment data, provide verification metrics, recognize certification documents, and acknowledge achievements. Could there be a common platform that would integrate each farmer's efforts, minimizing overlap and duplication? Could electronic platforms be developed that communicate with each other to reduce the cost of data management and streamline analytics and reporting? This will continue to be a work in progress for some time.

---

240 "Farm, Food, and Beyond - Our Commitment to Sustainability." p.2. Sustainable Farm & Food Initiative. (2015) www.sustainablefarms.ca/wp-content/uploads/2015/09/OurFarmSustainableAgenda-LR.pdf (Accessed May 30, 2017).

In summary, the SFFI is a "collaboration of Ontario's farm organizations and Provision Coalition representing Canadian food/beverage manufacturers. It provides a full scope, whole farm, whole value chain sustainability transparency system, focused on three pillars of sustainability: economic, social, and environment. Emphasis is on developing a pre-competitive data sharing platform, with the goal to provide a platform for harmonization of multiple tools and standards. It will help to manage risk to meet global demand for safe, healthy, and sustainable agri-food products." [241]

 What is the current Sustainable Farm and Food Initiative?

- A collaboration of Ontario's farm organizations and food/beverage processors
- Full scope, whole farm, whole value chain sustainability transparency system
- Fosters trust, transparency and mutual benefit
- Focused on three pillars of sustainability: **economic, social, and environment**
- Emphasis on developing a pre-competitive data sharing platform
- Help to manage risk to meet global demand for safe, healthy, and sustainable agri-food products

Figure 13.1 Partners for SFFI Development (Source: SFFI)

Some commodity groups are already out of the gate with specific sustainability programs to meet immediate market demands. Others have addressed some essential sustainability components that will eventually tie into the broader vision of consolidation and harmonization. Collaboration among partners at the national and international level, including appropriate government agencies, will be essential moving forward.

---

241   Sustainable Farm and Food Initiative Summary. https://www.sustainablefarms.ca/media-centre/ (Accessed February 20, 2018)

Extensive consultation has taken place as of 2017 with over fifty farm and food sector organizations. The findings are summarized below.[242]

Sustainable Farm and Food Initiative – What Do We Know?
- National and international sustainability landscape is complex and evolving rapidly;
- Ontario/Canadian farmers are facing a wide range of sustainability requests ranging from 'not on the radar to urgent';
- Certain commodity groups are pressing forward with their own strategies;
- This is leading to potential confusion in the marketplace, duplication of efforts and risk of missing a collaborative approach, leveraging funds, etc.;
- There is lack of consensus on the role of EFP/GYFP in a Canadian agriculture initiative;
- Stakeholders along the value chain do not want a piecemeal approach. Strongly support a cohesive, streamlined, national, whole farm, whole value chain solution.

Figure 13.2 SFFI – What do we know? (Source: SFFI)

In addition to this working group, I also participate, on behalf of OSCIA, in a national forum called the Canadian Roundtable for Sustainable Crops.

There is consensus that within Canada, the food industry has most of the sustainability requirements already in place, including strong regulations that address many areas of concern related to environment, labour, food safety, and social responsibility. We just haven't done a good job of connecting the dots, quantifying the achievements through verification, and telling the world about it.

In fact, a recent study by the Barilla Center for Food and Nutrition in Italy rates Canada at No. 3 overall on their food sustainability Index, and No. 2 behind Germany on agricultural sustainability.[243]

The SFFI working group commissioned Deloitte LLP to conduct a study entitled, *Environmental Farm Plan and Growing Your Farm Profits Gap Analysis*. The study compared the EFP and GYFP with ten other national and

---

242   Ibid.
243   "A Global Study on Nutrition, Agriculture, and Food Waste." Barilla Center for Food & Nutrition. www.foodsustainability.eiu.com/country-ranking (Accessed Sept. 6, 2017).

international sustainability programs based on twenty-five performance areas under three criteria: Planet (Environment), People (Social), and Profit (Economics).

> An over-arching goal is that a sustainability platform must be relevant to all of Canada and be recognized globally

Planet	People	Profit
• Water and wastewater management	• Food Safety	• Marketing
• Hazardous materials	• Animal and Poultry welfare	• Production
• Energy and climate	• Worker health and safety	• Financial Management
• Soil health	• Worker wellness	• Human Resources
• Nutrient management	• Worker training	• Succession Planning
• Pest management	• Worker housing/accommodation	• Business Structure
• Waste management	• Human rights	• Business Strategy
• Land Use	• Community engagement	
• Biodiversity and natural capital		
• Nuisances		

Figure 13.3 Performance areas to be managed, measured, verified and reported
(Source: Created by Deloitte LLP for the Sustainable Farm and Food Initiative)

The study found that, compared to international programs, EFP and GYFP provided a strong base for measuring Planet, People, and Profit criteria, with particular strength in Profit. Here's how we stacked up:

- Planet criteria: 76/100 topics were either equivalent or exceeded peers;
- People: 61/80 topics were either equivalent or exceeded peers;
- Profit: 68/70 topics were either equivalent or exceeded peers. [244]

At time of writing, Guelph-based Synthesis Agri-Food Network, Wilton Consulting Group, and Orion Global Business Sustainability Consultants are engaged in extensive consultations with stakeholder groups as part of the SFFI. The purpose of this consultation is to clarify needs and direction of a sustainability platform that will meet market requirements and streamline the assessment and

---

244 "Environmental Farm Plan and Growing your Farm Profits Gap Analysis." Deloitte LLP. www.sustainablefarms.ca/wp-content/uploads/2016/02/GAPsummary.pdf (Accessed June 2, 2017).

verification process. An overarching goal is that a sustainability platform must be relevant to all of Canada and be recognized globally. Equivalency analysis is underway using the Farm Sustainability Assessment (FSA, commonly known as FSA 2.0 tool) under the Sustainable Agriculture Initiative (SAI) Platform based in Brussels, Belgium.[245]

In moving forward, the SFFI is committed to sustainability for farmers, for the value chain, for government, and for civil society. Here is our commitment:[246]

For farmers?	For value chain?	For government?	For civil society?
• One data collection system • Trusted data management system |administrator • Seamless alignment with customer requirements • Opportunities for learning and continuous improvement	• One trusted and robust pre-competitive platform for Canadian agriculture • Seamless alignment with demands/requirements of other value chain stakeholders	• Sustainability data management system that supports export requirements, protects market share, builds public trust • Streamlined approach that optimizes government funding	• Ability to engage with farm and food sector to meet sustainability goals across the three pillars (social, environment, economic)

Figure 13.4 How does SFFI help? (Source: SFFI)

The EFP and GYFP are currently being benchmarked against the FSA. Industry leaders in Ontario feel the FSA is the most widely recognized and accepted farm assessment in the world, so if we can establish equivalency with the programs already in place, farmers will not have to duplicate their work. We would capitalize on twenty-five years of investment in programs such as

---

245   Sustainable Agriculture Initiative Platform. www.saiplatform.org/ (Accessed Sept. 8, 2017).

246   Sustainable Farm and Food Initiative. www.sustainablefarms.ca/wp-content/uploads/2017/09/SFFI_Summary_September2017.pdf (Accessed Sept. 26, 2017).

EFP. Wouldn't it be a major achievement if there could be a common language of sustainability that would allow us to communicate the successes of our Canadian agri-food system to the world? At time of writing, an internal equivalency assessment suggests that there are only a few additional questions (many referring to labour conditions) required for an EFP to meet the silver category of the FSA 2.0, and that the gold standard is achievable for many Canadian farms. We just have to prove it!

Canada has many of the components in place for sustainable food production to meet market demands from around the world. Strong leadership will continue to advance the agenda for consolidation and harmonization within Canada. How will a governance structure be assembled to solicit participation, encourage inclusiveness, inspire diversity, and provide verification to demonstrate the Canadian industry's commitment? Our next generation of competent leaders will no doubt rise to the challenge.

Conclusions

My thirty-year journey began as a result of strong scientific research concluding Great Lakes water quality was being degraded by significant amounts silt and phosphorus eroding from farmers' fields. There was growing awareness among astute farm leaders and government extension staff that conventional methods of plowing and tillage must give way to new soil and crop management systems. Ontario agriculture was evolving, and the concept of no-till was introduced into the province in the 1980s.

The challenges of Ontario's colder climate and generally wetter soils in the spring (compared to regions further south) was a barrier for no-till proponents to overcome. But practitioners soon recognized that the erosive forces of rainstorms and spring snow melt could be managed with high-quality soil drainage, and the adoption of coulter configurations for precise seed placement that left plant debris and straw on the soil surface. While previous generations had relied upon tillage and other mechanical weed control practices that disturbed the soil, new technology was being adopted that relied upon herbicides as an essential tool for weed control.

> Astute farmers soon learned that crop rotation and herbicide rotation were keys to perfect the no-till system and minimize the potential for herbicide resistance

Glyphosate became widely adopted because of its low impact on the environment, safety for users, and cost-effectiveness for improved soil management. Astute farmers soon learned that crop rotation and herbicide rotation were keys to perfect the no-till system and minimize the potential for herbicide resistance. Today, some are using strip tillage to combine the benefits of conventional and no-till techniques. Tilling thin strips in fields of row crops aerates and warms the soil to prepare it for seeding while leaving most of the field covered in plant residue that retains moisture and nourishes the soil. Organic producers are experimenting with roller-crimpers that flatten and desiccate green cover crops to form a thick mulch that protects and nourishes the next crop while eliminating or reducing the need for herbicides and reducing soil cultivation. More research is required to perfect the roller-crimper concept with respect to crop type, soil type, and timing for planting.

Research has also taught us that the benefits of no-till are limited if farmers just grow corn and soybeans in a rotation. Adding winter wheat to the rotation opens the door to using green manure crops that enhance biological activity, especially if manure or compost are added to the mix. Better yet, the queen of forages, alfalfa, would be the ultimate and OSCIA celebrates an Ontario Forage Master each year to focus on exceptional forage managers.

However, while no-till and conservation tillage are important ingredients, soil protection is only one component of sustainable farm management. Many other topics needed attention, ranging from safe drinking water and the impacts of bacteria and nitrates on water quality; petroleum storage; pesticide management; nutrient management plans; and best practices for managing other resources such as woodlots, wetlands, and water courses adjacent to farms. The Environmental Farm Plan (EFP) program emerged as stakeholder groups recognized the need to develop tools for the comprehensive assessment of on-farm activities, best practices and action plans to improve environmental conditions. Ontario's user-friendly approach caught attention of practitioners around the world. The key to EFP's success in Ontario was the strong synergy achieved by combining government expertise, the political drive of farm leaders, and grassroots delivery by OSCIA's network of farmers. Similar collaborations focused on key issues such as nutrient planning and water-quality protection.

There was no point in spending millions of dollars and endless hours on managing and improving soil without also developing a solid business model that incorporates a succession plan for future generations. To complete the sustainable agriculture formula, Growing Your Farm Profits (GYFP) evolved to identify best practices to manage the economic and social aspects of farming: marketing, production, financial management, human resources, social responsibility, succession planning, business structure, business strategy.

The Sustainable Farm and Food Initiative (SFFI) brought industry leaders together to capitalize on progress to date on the environment, social, and economic pillars of sustainability to help position Canada's food system as a world leader.

> This is a callout to respect planet, people and profit as the overriding pillars for agricultural sustainability

The task of fixing soil is far from done. The 2017 Summit on Canadian Soil Health flagged continuing concerns. Soil, water, climate change, and food production are so closely interrelated that future stewards of the land must integrate all four of these components into future business plans to accomplish what some previous civilizations could not achieve—a sustainable society. This has been a callout to respect planet, people and profit as the overriding pillars for agricultural sustainability.

In parallel to support soil health, the Ontario government has recently taken a bold step by announcing New Horizons: Ontario's Agricultural Soil Health and Conservation Strategy, recognizing that healthy soils are key to our $13-billion food economy and 800,000 jobs.[247]

This book has focused on high-profile programs at OSCIA that provide awareness, training, and cost-share for continuous improvement to meet the goals of sustainable agriculture. Soil management is a key component and a top priority for OSCIA and its 4,000-member network.

The contributions of OSCIA's eleven regional associations encompassing fifty-three counties/districts across Ontario is truly where the rubber hits the road. Each year there are hundreds of field days, workshops, and farm tours to investigate new methods, showcase innovative ideas, and perfect or tweak

---

247 OMAFRA. New Horizons: Ontario's Agricultural Soil Health and Conservation Strategy. http://www.omafra.gov.on.ca/english/landuse/soil-strategy.pdf (Accessed May 10, 2018).

some annoyance that is hindering farm efficiency. New ideas spring forth at every meeting and there could be a whole book written on the grassroots activities of OSCIA's membership, rich in culture, camaraderie, and generous hospitality. At the provincial office, we simply provide the locals with infrastructure and tools to help tackle their priorities. As indicated previously, professional staff from OMAFRA, especially the Field Crops Unit, play a crucial role in providing oversight, designing protocols, conducting analysis and reporting results of on-farm research. A sequel to *The Soil Fixers* could be assembled to capture the many local dynamics and successes. Thankfully, the *Crop Advances* publication summarizes plot results each year.

I've touched on the critical role of science in providing the tools to improve the genetics of seeds, enhance plant nutrition, and ensure bountiful harvests by protecting crops from disease, insects, and weeds. Agri-business is an important partner, too, in providing expertise to guide the selection of crop inputs and contribute to fixing soil. Most of the agronomists working with farmers are trained, tested and certified through the Certified Crop Advisor Association (CCA), which has close to 650 members in Ontario.[248]

To support the expanding organic industry, there are now nineteen organizations accredited by the Canadian Food Inspection Agency (CFIA) to certify food products as organic. Nine of these accredited organizations are Canadian. The other ten outside of Canada would comply with Canadian organic regulations to facilitate trade with Canada.[249]

Gene editing is an exciting new tool that promises to solve a variety of problems including reducing or eliminating pesticide use, snuffing out allergens in food, improving nutrition, and enhancing the flavour of food. It may also assist medical science in eliminating some human and animal diseases. What if alfalfa, the ultimate plant for soil protection and soil fixing—and a great source of protein for livestock—could have a few genes tweaked to

---

248 Ontario Certified Crop Advisor Association. http://ccaontario.com/ (Accessed April 6, 2018).

249 Canadian Food Inspection Agency. Certification Bodies accredited by the CFIA – in Canada, http://www.inspection.gc.ca/food/organic-products/certification-and-verification/certification-bodies/in Canada/eng/1327861534754/1327861629954 (Accessed January 31, 2018).

make it one of the tastiest and most nutritious salads for humans on this planet! Wouldn't that change our cropping opportunities and contribute to sustainable agriculture!

Although gene engineering has safely led to new efficiencies in food production and especially in conserving and improving soil, each new development must be examined and considered on its own merits. Thankfully, in Canada each application to register a new seed variety is closely scrutinized by exceptionally trained scientists governed by a vigorous regulatory process.

There is little reference in *The Soil Fixers* to land use planning. Although the topic is of interest to me personally, it's been left to other organizations to address agriculture's interests related to smart growth, greenbelt protection and other aspects of the Ontario government's Provincial Policy Statement. OSCIA is focused on addressing land-management issues faced by farmers and providing the tools to help them be better food and fiber producers. My thirty-two-year journey has resulted in many reflections on how Ontario society is managing our landscape. Further, when I stand at the grocery checkout and see some of the poor choices (read non-healthy) that fill the shopping carts of my fellow shoppers, I often wonder whether they even care about our efforts aimed at achieving sustainable food production.

Score: Horse Farms 12, Veggie Markets 1

Nothing brings the topic of rural planning to the forefront more vividly than my many road trips throughout Ontario. Even on the commute between our farmstead at New Hamburg and the OSCIA office in Guelph, I pass a dozen horse farms but only one roadside vegetable stand. Why isn't locally grown food more readily available and easily accessible in our community? Why aren't vegetable farms as commonplace as horse farms? Over thirty years, I've clocked close to 600,000 kilometers to and from work, so I've had lots of time to ponder!

I have nothing against horses or horse farms. In fact, I enjoy watching horses compete at shows and wouldn't object to one or a few in a fenced paddock in view of my porch. I even volunteered for a couple of years at a therapeutic riding school. The landscape on most horse farms is hay and pasture, crops that stabilize and improve the soil. OSCIA programs have assisted many horse farms over the years and many of them have participated in the EFP program.

I would love to support food from local producers, but throughout my career have had little time to drive out of my way to go to farmers' markets or roadside stands. Why isn't growing produce more of a priority for landowners adjacent to urban areas (and potential markets). After all, there is a significant population in the Kitchener-Waterloo, Cambridge, and Guelph region.

Could the answer be that growing produce on these smaller tracts of land generates low profit margins, is incredibly labour intensive and requires long hours over the summer months compared to other commodities, and that readily available jobs in the city are more stable (and lucrative) for a secure family income? Would this be a strong indicator that produce farming in our communities is less sustainable? Could this area support more Community Supported Agriculture (CSA) farms? I have a few family friends who operate CSA farms and they work incredibly hard, with a level of intensity and fortitude not many could tackle.

What will be OSCIA's future role in developing policy for land use planning, farmland preservation, small holder farmers, food security, and local food? Does OSCIA have a role in transferring Canadian knowledge and skills to countries with more traditional or subsistence-level agricultural systems? OSCIA's current mandate does not actively support international development.

As I ponder these questions, I realize that sustainability has so many touchpoints, so many external influences and ideological pressures, that it is easy to miss the mark. But by bestowing control to those who own and manage the land, and equipping them with the resources and tools that support continual improvement, we also give hope to future farmers and help our industry inch ever closer to the target—sustainability. Our food industry has given Canadians exceptional food quality and a multitude of choice at the lowest cost in our history. What will it take to ensure that the rest of the world's population can enjoy this similar advantage?

I have no doubt the next thirty years at OSCIA will not mirror the past thirty years. Many of the challenges outlined in OSCIA's past journey have only recently begun to take root and will undoubtedly affect OSCIA's priorities for the future. In 2019, OSCIA will celebrate its eightieth anniversary. Its mandate continues to be as relevant today as it was in 1939, perhaps even more so. My involvement has been serendipitous, and I have been fortunate to have worked with such wonderful colleagues.

In meeting farm families and networking with the next generation of producers, I see optimism. The questions surrounding sustainability, markets and the environment will, I'm sure, be a priority for future projects at OSCIA. Undoubtedly, OSCIA members, directors, executive leaders, skilled staff, and partners will provide the leadership for training, communication, research, and cost-sharing of innovative technology. Tomorrow's leaders will assist in the adaptation of as-yet unheard of technology, by an ever-evolving industry. Our leaders in 1939 could not possibly have foreseen the incredible benefits enjoyed by society today, thanks to the efforts of the OSCIA and its stakeholders and partners. I can't wait to see what unfolds in the next thirty years!

Acknowledgements

Authoring a book does not happen by accident. I am most grateful to the executive and board of the Ontario Soil and Crop Improvement Association (OSCIA) for giving me the green light to take on the task of capturing highlights of my thirty-plus-year career. Thanks also to Andrew Graham, OSCIA's Executive Director, who kindly provided flexibility of time to record my journey.

Fact checking was made much easier, especially for the 1980s and 1990s, with many conservation program details captured and readily available through a website developed by Dr. Bruce Bowman, a retired scientist from Agriculture and Agri-Food Canada: http://agrienvarchive.ca/pubs/btb_cv.html

Dr. Bowman was intimately engaged in soil and water conservation research through this era and he made it his mission to catalogue the multitude of activities. Thank you, Bruce!

I appreciate very much the reviewers and their constructive comments from my earlier drafts: Keith Reid (Soil Scientist, AAFC); H. J. Smith (Environmental Management Specialist OMAFRA); Margaret Stewart (Office Manager, OSCIA); Andrew Graham (Executive Director, OSCIA); Gord Green (2016 OSCIA President); Mack Emiry (2017 OSCIA President); Dr. Wayne Caldwell (Professor, School of Environmental Design and Rural Development, University of Guelph); Dr. Rebecca Moore (Manager, OMAFRA-U. of Guelph Communication, Office of Research, University of Guelph); David Armitage (Director of Regulatory Reform, Ontario Federation of Agriculture); and many other fact-checkers. You know who you are.

Thanks also to my wordsmith and editor, Barry Gunn. There comes a time when the creative well dries up and it was great to turn the manuscript over to someone of Barry's creative talents. I am most appreciative of the generous support from Dan Needles, author, playwright, and humourist, in writing the Foreword. Mr. Needles is well known among our members, often entertaining as a guest speaker. His insights are interesting and compelling. Contained within the covers of The Soil Fixers are additional endorsements. I highly value comments from these highly respected scientists and colleagues. Sketch artist Warren Muzak, Waterloo, Ontario provided exactly the image I was seeking for the book cover, as it reflects the multitude of field days, on-farm tours, and experimental plots where farmers and students of the land seek greater understanding of soils from scientists and extension experts.

In reflecting back over thirty-plus years, I would not have survived were it not for the support of hundreds of colleagues, administrators, and county/district committees across Ontario. Many of the new programs and tasks we took on ventured into uncharted territory, perhaps even with some elevated administrative risk. But with a positive attitude, creative approaches, and exceptional dedication of staff, we were able to record many achievements at OSCIA.

OSCIA program activities would not have happened were it not for the strong and enduring support of our government partners, Ontario Ministry of Agriculture, Food and Rural Affairs, and the federal counterpart, Agriculture and Agri-Food Canada. Private-public partnerships have done well in Ontario. Industry leaders, particularly with the formation of the Ontario Farm Environmental Coalition, provided a new focus on environmental responsibility and a vision for how Environmental Farm Plans (EFP) could elevate our on-farm adoption of best practices. The task of mastering sustainable farms is not yet complete. It is a continual improvement process. Our innovators and leader will continue to advance the goal posts.

Last but certainly not least, I wish to thank my many family members who have been most supportive in my career journey and my ambitions to document highlights in The Soil Fixers. When selecting just the right word or phrase, Sandra's literary talents helped me out with a most suitable fit.

The Real Meaning of Sustainable Farms and Food

A new list of characters will emerge at OSCIA over the next thirty years. I am confident they will be equally inspiring stewards of our land. Someone else, however, will have to take up the authorship to record their inspiring stories.

Appendix 1
OSCIA Projects and Programs Listing

(1980s – 2018)

1983	Co-sponsored "Soil Today - Food Tomorrow" - December 1983.
1984	Presented a brief to Standing Senate Committee on Agriculture, Fisheries and Forestry - May 1984.
1985	Organized "Improving the System - Agricultural and Drainage in Ontario" Forum - December 1985.
1986	Co-sponsored "Forage Your Key to Profit" Seminars - Winter 1986.
1985 - 1990	Implemented the "Soil Conservation Award" in cooperation with OMAF, 1985 - 1990.
1987	Presented brief to Crop Insurance Commission - March 1987.
1987	As a member of the Ontario Agricultural Crop Protection Committee, presented a brief on Grower Certification - June 1987.
1987 - 1988	Field Days, lead by Ciba-Geigy Executive Tom Sawyer at Honeywood Research Farm, Plattsville, ON, July '87-'88
1987	Acted as a 'clearing house' for local concerns in the form of resolutions at our annual meeting, which are then referred for action to various agencies. Many of these concerns involves production issues, regulations, and need for technical assistance.
1987	Encouraged county associations to develop soil conservation demonstration projects, local educational programs.
1987	Support OMAF programs - e.g. the Soil Conservation and Environmental Protection Assistance Program announced in 1983.
1987	Presented a brief on a proposal to implement and deliver the Land Stewardship Program - June 1987.
1988	Signed contract with OMAF, January 1988, for the County/District Program Delivery and Service section of the Ontario Land Stewardship Program.
1988	Co-sponsored "Conservation Farming '88," held at the Woodstock Research Station - June 1988.
1988	Co-operated with the Tillage 2000 and Side-by-Side Trials co-ordinated by OMAF and University of Guelph.
1988	Co-operated with the Weather Recording Program.

1988	Developed and implemented the Ontario Forage Master Competition - 1988 - present.
1990	Signed contract with Agriculture and Agri-Food Canada, September 1990, to deliver the Permanent Cover Component of the Nation Soil Conservation Program for Ontario Region.
1991	Signed contract with OMAF, January 1991, for the County/District Program Delivery and Service section of Land Stewardship II Program.
1991	Prepared and distributed the "Ontario Grown Food" poster with assistance from Foodland Ontario - April 1991.
1991	Signed contract with Agriculture and Agri-Food Canada, October 1991, to administer the "Ontario Farm Groundwater Quality Survey."
1992	Signed contract with Agriculture and Agri-Food Canada, Spring 1992, to deliver the High Crop Residue Program.
1992	Major participant in helping to organize "Rural Routes '92."
1992	Co-ordinated Water Well Management Demonstration Project supported by AAFC through Agricultural Adaptation Council.
1992	Signed contract with Agriculture and Agri-Food Canada, Summer 1992, to deliver the Permanent Cover II Program.
1993	Signed contract with Ontario Farm Environmental Coalition, December 1993, to serve as the delivery agent for the Environmental Farm Plan Program. Continues through 2008.
1994	Began agreements with Canada's Outdoor Farm Show to sponsor the shows. Agreements signed for 1994-1996, 1997-1999, 2000-2002, 2003-2005, 2006-2008 and continues to present.
1995	Co-ordinated Wetlands, Woodlands Wildlife Program for Environment Canada - Canadian Wildlife Service.
1997	Co-ordinated Farm Fuel Storage Demonstration Site and prepared associated brochure "Spill Containment for Farm Fuel Storages."
1997	Assisted with "Crop Biotechnology for a New Millennium: conference series held at four location across Ontario - December 1997.
1997	OSCIA Major/Project Grants - 1997 - ongoing - assumed responsibility for administration of these grant programs. Manager, Crop Technology signs off on approved grants for OMAF.
1995-1999	Great Lakes Basin Comprehensive Farm Planning Network, funded by the US-based Great Lakes Protection Fund.
1998	Co-ordinated Restoration of the American Chestnut - Farm Response to a Species at Risk.
1998	Co-ordinated the investigation Wildlife Impact Assessment for Ontario Agriculture.

1998	Assisted with 1998/1999 DEKALB Roundup Ready Showcase.
1998	Launched OSCIA Website - March 1998 - www.ontariosoilcrop.org
1999	OSCIA Special Education Programs 1999 - new grant to assist active organizations further their impact of education events/activities for members/communities.
1999	Ontario Field Trails Online - OFTO, an electronic database designed for performance trials, was established 1999. The OFTO database standardized the data collection for corn, wheat, and soybean trials conducted on Ontario farms. The ability to compare results from across the province as plots are harvested gave farmers valuable information to better select hybrids/varieties for their operations. - 1999-2001.
1999	OSCIA Executive Workshops - ongoing - the OSCIA Executive meets with local associations on a continuing basis.
1999	OASCC Committees - ongoing - OSCIA appoints its directors to represent grower input on many of the OASCC committees.
1999	Gate Signs - OSCIA introduced new member gate signs.
2000	OSCIA Regional Associations - reorganization of province into 11 regional associations. Development of OSCIA Regional Partner Grants and OSCIA Regional Communication Grants to support activity within the regional association structure.
2000	OSCIA list server - News and Views forwards one to three articles weekly on crop production and soil management-related topics via email to subscribers. The list server also features a comprehensive "Calendar of Events" with upcoming soil and crop-related meetings, events, field days, and conferences.
2001	Assisted in coordinating Riparian Zone Workshop in Cambridge.
2002	Prepared and released Wildlife Wise publication that featured on-farm efforts to manage problem wildlife and species at risk.
2002 - 2005	ICAT (Independent Corn Adaptability Trials) - The objective of ICAT was to establish a series of independent, local-level co-ordinated and organized corn strip trials. ICAT offered members a powerful tool to select superior corn hybrids that perform best on their farms to increase yields.
2002 - 2005	ICAT Grant - Introduction of grant to assist with the co-ordination of planting, in-season plot assessments, and harvest data collection and forwarding of data to a central data entry for Ontario Field Trials Online.
2003 - 2006	Ontario Greenhouse Gas Mitigation Project - (Soil Conservation Council of Canada and AAFC) - To profile new or emerging BMPs that are effective at reducing or removing GHG from the atmosphere.
2004	Signed agreement with OMAFRA to deliver the Nutrient Management Financial Assistance Program (2004-2006).

2004	Nutrient Management Planning Training Courses Co-ordination - Provided co-ordination support to OMAFRA for the Nutrient Management Certification Training Courses.
2004	Prepared and released Probing Problem Wildlife as part of Wildlife Action Project supported by Agricultural Adaptation Council.
2005	Signed agreement with OFEC to deliver EFP and Canada-Ontario Farm Stewardship Program (2005-2008).
2005	Signed agreement with AFFC to deliver Greencover Canada (2005-2008).
2005	Signed agreement with AAFC to deliver Tier 1 of Canada Ontario Water Supply Expansion Program (2005-2008)
2005 - 2007	On-farm Evaluation of Biodiesel (Natural Resources Canada) - evaluated practical on-farm problems associated with biodiesel use on different farm operations.
2005 - 2008	On-farm Evaluation of Optical Sensor Technology to Minimize Environmental Impacts and Maximize Production Efficiencies Associated with N Application in Corn (Environmental Technology Assessment for Agriculture Program, AAFC).
2006	Signed agreement with Oak Ridges Moraine Foundation to deliver the Oak Ridges Moraine Environmental Enhancement Program (2006-2008).
2006	Signed agreement with Friends of the Greenbelt Foundation to deliver the Greenbelt Farm Stewardship Program (2006-2008).
2006	Assisted with delivery of environmental conference on Understanding What Compels Producers to Adopt Beneficial Management Practices, Mississauga.
2006 - 2009	Management of Agricultural Landscapes with Wetlands and Riparian Zones: Economic and Greenhouse Gas Implications (Ducks Unlimited Canada - ACAAF) - address gaps in understanding of the environmental role and economic value of wetlands and riparian zones in agricultural landscapes across Canada.
2006	Producer Consultation Meetings: Clarifying No-Till Practices (AAFC) - in conjunction with a national series to help better understand producer/industry issues related to the adoption and implementation of zero tillage and other beneficial soil management practices.
2006 - 2008	Evaluating Cover Crops' Ability to Carry Manure (Canada-Ontario Research and Development Program - Agriculture Adaptation Council). This project addressed the possibility that where manure is applied to Ontario fields in the summer or early fall, cover crops such as oats could sequester nitrogen (N) and relay it into the corn crop the following year on 10 farms across Ontario.
2006 - 2008	Mark Cullen, noted garden writer, engaged to promote stewardship techniques used by farmers that could be applied by gardeners, including newspaper columns, radio broadcasts and very popular brochure.

2006 - 2010	Partners Validating Agronomic Traits (PVAT) (Ontario Agri-business Association from the Fertilizer Institute of Ontario Fund). Addresses Agriculture's role in adopting fertilizer BMPs which have potential to reduce the impact of commercial nutrients on the environment, and the opportunity for Ontario producers to supply for new niche markets by targeting nutrient application to meet a new crop quality in demand.
2007	Signed agreement with OFEC to deliver the Ontario Ministry of the Environment's Source Protection Program (2007-2008).
2007	Worked alongside OMAFRA with support from Agricultural Adaptation Council and the Agricultural Training Institute to compile and introduce Growing Your Farm Profits self-assessment for farm businesses.
2007	Conducted an Agricultural Land Value Survey for the Drinking Water Program Management Branch of the Ontario Ministry of the Environment.
2008 - 2011	Nutrient Management BMP Demonstration Grant and Regional Outreach Grant - OMAFRA funding to support new communication activities of regional SCIAs that promote the adoption of Nutrient Management BMPs to the non-regulated NM Act.
2008-2012	Lake Simcoe Agricultural Stewardship Program provided environmental cost-share funding for farmers in the Lake Simcoe Watershed to implement best management practices.
2008 - 2010	Ontario Red Clover Research (Canada-Ontario Research and Development Program - Agriculture Adaptation Council). Continued research regarding the benefits of growing red clover.
2008 - 2010	Foliar Product Impact on Seed Quality (Ontario Seed Growers' Association) - verify if foliar products used without pre-existing conditions of insect of disease pressure increase seed quality and yield, and highlight the findings at Canada's Outdoor Farm Show, Woodstock.
2008 - 2013	Ontario Drinking Water Stewardship Program, funded through the Early Actions component of the Ontario Ministry of the Environment (through the Ontario Federation of Agriculture), provided financial assistance for farmers in source water protection areas.
2009	Signed Agreement with Ontario Farm Environmental Coalition to renew delivery of EFP and the associated Canada-Ontario Farm Stewardship Program.
2009	Signed Agreement with OMAF to renew delivery of Growing Your Farm Profits and the associated Business Development for Farm Businesses.
2009	"Field-Scale Agricultural Biomass Research and Development Project" was launched by OSCIA and partners with support from AAFC through the Canadian Agricultural on-Adaptation Program, delivered in Ontario by the Agricultural Adaptation Council.
2009 - 2012	Lake Simcoe Farm Stewardship Program provided environmental cost share funding to farmers in the Lake Simcoe watershed to adopt best management practices. Funded in part by both provincial and federal governments.

2010	Worked with partners to launch the four-year "Superfoods for Health - Amaranth and Quinoa Production in Ontario."
2011	EnvironMERIT - A Conservation Tender concept developed for riparian systems by OSCIA, marks a turning point in cost-share program design.
2012	OSCIA partnered with Trent University, with funding from MNR to conduct on-farm research in Renfrew County, to investigate whether intensive rotational grazing systems can boost quality of habitat for grassland birds such as Bobolink.
2012	The "Grassland Habitat Farm Incentive Program" was introduced to encourage producers to adopt grassland management practices that support production and create habitat for wildlife species at risk.
2012	OSCIA worked with OFEC to compile and release "EFP: Right Actions in the Right Places - Environmental stewardship through responsible agricultural nutrient management."
2012	OSCIA funded the report "A Safe Harbour Policy for Canada?: Examining the potential for safe harbour agreements within the confines of the federal Species at Risk Act."
2012	Through EFP, a spatial analysis of nutrient management Best Management Practices from April 2005 to March 2010 prepared by an analyst from Agriculture and Agri-Food Canada.
2013	With funding from the Friends of the Greenbelt Foundation, OSCIA delivered "Farming Power - Energy Innovation in the Greenbelt." This merit-based program offered cost-share for energy-saving retrofits.
2013	For the sixth consecutive year, OSCIA delivered the "Species At Risk Farm Incentive Program" with support from Ministry of Natural Resources and Environment Canada.
2013	"Potential Role of Ontario Environmental Farm Plan in Responding to Sustainability Demands of the Agri-food Supply Chain" was co-ordinated by OSCIA and other partners and carried out by the George Morris Centre with support from Agriculture and Agri-Food Canada through the Canadian Agricultural Adaptation Program, delivered in Ontario by the Agricultural Adaptation Council.
2013	Coordinated development and registration for the initial delivery of the Advanced Farm Manager Program, for farmers who had participated in the Growing Your Farm Profits workshop to receive advanced educational opportunities.
2013	OSCIA selected by OMAF to deliver the five-year Growing Forward 2 educational and funding assistance programs to Ontario producers.
2013 - 2015	OSCIA delivered the Grassland Habitat Farm Incentive Program, funded by the Habitat Stewardship program of Environment Canada to support farmers in protecting grassland bird habitat.
2014	Delivered the Porcine Epidemic Diarrhea (PED) Biosecurity Special Intake to the swine sector as part of GF2 activity.

2014	An on-farm research project was conducted with Agriculture and Agri-Food Canada (Dr. Dan Reynolds, Harrow) to better quantify the physical characteristics of healthy soil. The project was funded by the Water Resource and Adaptation Management Initiative provided by OMAFRA and delivered by Farm and Food Care in addition to the Farm Innovation Program of Growing Forward 2, delivered by the Agriculture Adaptation Council.
2014	OSCIA members participated in on-farm corn trials for a neonicotinoid seed treatment efficacy study.
2014	Water's Edge Transformation (WET) Program for stewardship projects in agricultural riparian systems in Lake Simcoe, Severn Sound and Nottawasaga watersheds was funded by Ontario Ministry of Agriculture, with technical assistance provided by conservation authorities, based on a competitive bid process.
2015	With support from Environment Canada and the Ministry of Natural Resources and Forestry OSCIA delivered the Species at Risk Farm Incentive Program to support species at risk.
2015	OSCIA launched a four-year agreement to deliver the Great Lakes Agricultural Stewardship Initiative (GLASI) with support from the Ontario Ministry of Agricultural, Food, and Rural Affairs and the Government of Canada to address phosphorus issues in the great lakes. Major components were the Farmland Health Check-Up, the Farmland Health Incentive Program and the Priority Subwatershed Project.
2015	The OSCIA Soil and Crop Sustainability Fund was established in partnership with the University of Guelph. It was launched at the February AGM with many past-presidents contributing as founding partners. With additional funds provided by OSCIA friends and the board, the fund quickly expanded to $180,000.
2015	OSCIA participated on the Field Crops Research Infrastructure and Co-ordination along with OMAFRA, University of Guelph and the Grain Farmers of Ontario to consider ways of extending the investment of Ontario research farms into a Network of Excellence.
2015	OSCIA has been part of an industry committee to co-ordinate a sustainability agenda, the Sustainable Farm and Food Initiative (SFFI) to respond to growing demands by the value chain to verify sustainable farming and food production practices.
2016 - 2018	SARPAL (Species At Risk – Partnerships on Agricultural Lands) is an Environment and Climate Change Canada initiative that is focused on working with farmers to support the recovery of species at risk on agricultural land, focusing on bobolink and the American badger. Components are the Grassland Stewardship Program, Badger Way, and education and outreach activities.
2017	Grasslander, an on-going citizen science project provides an online platform for Ontario farmers to contribute data to scientific grassland bird conservation efforts, recording sightings of the eastern meadowlark and boblink.
2018	Canadian Agricultural Partnership – five-year commitment by AAFC and OMAFRA for many project opportunities

Appendix 2

OSCIA Past-Presidents with Year Served and Home County/District

2018	Peter McLaren	Lanark
2017	Mack Emiry	Sudbury District
2016	Gord Green	Oxford
2015	Alan Kruszel	Stormont
2014	Allan Mol	Thunder Bay District
2013	Henry Denotter	Essex
2012	Joan McKinlay	Grey
2011	Max Kaiser	Lennox & Addington
2010	William Barry Hill	Brant
2009	Murray Cochrane	Algoma
2008	Pat Lee	Oxford
2007	Frank Hoftyzer	Peterborough
2006	Keith Black	Huron
2005	Kevin Ferguson	Ottawa-Carleton
2004	Steven Eastep	Wellington
2003	Lloyd Crowe	Prince Edward
2002	Greg Kitching	Halton
2001	Fred Judd	Norfolk
2000	Ben Kamphof	Thunder Bay District
1999	Allan Yungblut	Niagara North
1998	Denis Perrault	Russell
1997	Jim Fischer	Bruce
1996	Jim McWilliam	Durham West
1995	Allan Brown	Simcoe North
1994	Victor Roland	Perth
1993	Ken McCurdy	Hastings
1992	Elwin Vince	Kent
1991	Maurice Martin	Elgin
1990	Jim Yungblut	Niagara South

1989	Bill Zandbergen	Dundas
1988	Don Hill	Grey
1987	Richard Sovereign	Halton
1986	Donald MacDonald	Frontenac
1985	Frank Little	Essex
1984	Laurence Taylor	Huron
1983	Grant Richardson	Haldimand
1982	Graydon Bowman	Temiskaming
1981	Ken Patterson	Middlesex
1980	John Noel Dessaint	Ottawa-Carleton
1979	John McGill	Lanark
1978	John Benham	Wellington
1977	George Gardhouse	Peel
1976	Howard Huctwith	Lambton
1975	James Barrie	Waterloo
1974	Leland Wannamaker	Lennox & Addington
1973	Russell Morrison	Durham East
1972	Alfred Baudette	Stormont
1971	Henry Davis	South Simcoe
1970	Ross Leedham	Norfolk
1969	Howard Salmon	Wentworth
1968	Fred Cohoe	Oxford
1967	Reg McCann	Northumberland
1966	Leonard Trivers	Algoma
1965	Robert Sparrow	Ottawa-Carleton
1964	George Stuffel	Dundas
1963	Eugene Lemon	York
1962	Hugh Glasgow	Kent
1961	Grover Smith	Prince Edward
1960	Morris Darby	North Simcoe
1959	George van Sickle	Brant
1958	H.H.G. Strang	Huron
1957	Jas A. McBain	Elgin
1956	W.W. Dawson	Peterborough
1955	Andy Johnson	Renfrew

1954	W.J. Schneller	Waterloo
1953	Howard Harper	Ontario
1952	William Wallace	Essex
1951	J.B. Graham	Wentworth
1950	L.B. Mehlenbacher	Haldimand
1949	H.H. Mcnish	Leeds
1948	A.A. McTavish	Bruce
1947	T.A. Wilson	Lanark
1946	Frank V. Dedrick	Norfolk
1945	G.R. Richard	Durham East
1944	Gordon Hancock	Peterborough
1943	H.E. Simpson	Simcoe North
1942	W.E. Breckon	Halton
1941	R.J. McCormick	Brant
1940	Clarke Young	York
1939-1938	Alex. M. Stewart	Middlesex

Honorary Presidents and Year Served

2018	Dawn Pate
2017	Gerald Beaudry
2016	Peter Johnson
2015	Dr. David Hume
2014	Jim Arnold
2013	Colin Reesor
2012	Lee Weber
2011	Peter Hannam
2010	Don Lobb
2009	Lyle Vanclief
2008	Ginty Jocius
2007	Bill Parks

2006	Dr. Allan Hamill
2005	William Curnoe
2004	Harvey Wright
2003	Dr. Terry Daynard
2002	Ken Knox
2001	Dr. Charles S. Baldwin
2000	Dr. Gordon Surgeoner
1999	Dr. Rob McLaughlin
1998	Galen Driver
1997	Ralph Shaw
1996	John Benham
1995	Vernon Spencer
1994	Everett Biggs
1993	Jack Riddell
1992	Tom Sawyer
1991	Dr. Stan Young
1990	Dr. Clay Switzer
1989	William A. Stewart
1987	Vacant
1986	Mike Clitherow

Appendix 3

CONFIDENTIALITY AGREEMENT FOR ENVIRONMENTAL FARM PLANS

(Source: OSCIA Archives)

| Minister
Ministre | Ministry of
Environment
and Energy | Ministère de
l'Environnement
et de l'Énergie | 135 St. Clair Avenue West
Suite 100
Toronto ON M4V 1P5 | 135, avenue St. Clair ouest
Bureau 100
Toronto ON M4V 1P5 |

DEC 1 8 1995

RECEIVED
JAN - 4 1996

Mr. Gord Surgeoner
Dept. of Environmental Biology
University of Guelph
Guelph, Ontario
N1G 2W1

Dear Mr. Surgeoner:

I am very pleased to provide you with a copy of our new *Policy and Guideline on Access to Environmental Evaluations*. This innovative and important policy has been developed through the efforts of my Ministry, the Advisory Committee on Use of Environmental Information, and representatives from business, the farm community, and environmental groups.

The purpose of the policy is to encourage Ontario businesses to use environmental evaluations by ensuring their confidentiality. The policy clarifies the circumstances under which the Ministry of Environment and Energy (MOEE) may request access to environmental evaluations. It also outlines the steps which MOEE staff must follow in securing access to these evaluations during abatement inspections, investigations, and emergency situations. Further, the policy provides provisions for demonstrating "good faith" and encourages the use of "program approvals" once a problem has been identified.

This policy has generated tremendous interest. I want to thank all of you who participated, through the Advisory Committee, various stakeholders' meetings, and written comments. Your input has been invaluable in producing this policy. I believe that we have succeeded in striking the appropriate balance between the exercise of MOEE's regulatory responsibilities and the right of individuals to assess their own environmental performance without fear of self-incrimination.

Page 2

I am confident that this policy will promote environmental protection while ensuring continued effective enforcement. In issuing this document, I would like to challenge business, to demonstrate its use of environmental evaluations as a standard practice - one which is both good business management and good environmental management.

Yours sincerely

Brenda Elliott
Minister

Enclosure

Appendix 4

REPORT TO THE ONTARIO FEDERATION OF AGRICULTURE EXECUTIVE AND ENVIRONMENTAL COMMITTEES

October 18, 1996
By David Armitage
Director, Regulatory Reform

"The same federal regulation (Surface Water Treatment Rule) required filtration of surface water used to supply a municipal population (applied to both Skaneateles Lake for the City of Syracuse and New York City watersheds). New York City, with a population of nine million, would have required a filtration plant costing $6-$8 billion to build. Operating expenses would have been staggering at approximately $1 million per day. They have managed to avoid the cost of a filtration plant by implementing a program of comprehensive farm planning aimed at the 450 farmers in the watersheds that New York City draws its water from. Specifically, the watersheds include Catskill, Delaware and Croton. Together these watersheds comprise a system of reservoirs, lakes, holding tanks, aqueducts, tunnels and rivers having a combined surface area of 5,200 square kilometres. The storage capacity is approximately 548 billion gallons of water, with New York City drawing approximately 1.5 billion gallons per day.

A pilot project dealt with ten farms in the watershed. Following the pilot, a $35.2 million agreement was written to underwrite the cost of capital improvements associated with implementing BMPs. Fund are provided over a five -year period. On average, $75,000 is allocated per farm in the program, with a maximum capital contribution of $100,000.

The program has been characterized as a Human Resource Management Plan rather than a Water Resource Management Plan, given its focus on BMPs.

Comments

Those on the tour were generally agreed that there were several concerns associated with both programs.

First, while farmers were involved, they were involved only as individual land owners. There was no evidence of farm organization involvement in the development and delivery of the program. This may be one reason that farmers in the New York City watersheds seemed reluctant to participate in the program despite the financial benefits. Hesitancy on the part of these farmers was also a function of historic inertia dating back to the displacement of rural residents when the New York City reservoir system was established in the first half of this century.

Second, the strategy of providing 100 per cent funding was hard to justify. There is little or no motivation on the farmers' part when a 'team of experts' with access to high levels of funding make recommendations. On one farm that was visited, the program had not only covered the entire cost of designing and installing an earthen manure storage system, but had also purchased a manure pump and a manure spreader. Some wondered if the owner might also be looking to the program to purchase a tractor capable of pulling the manure spreader, since there did not appear to be a tractor of sufficient horsepower on the farm.

Third, there were situations where the programs were investing in a farm that might reasonably have been purchased and taken out of production, given the dollars involved. The farm referenced above would fit this category, as would a dairy we visited that was virtually in the centre of a small town. The owner of the dairy was considering expanding his cow herd from 108 milking to 300. One of the problems that this individual had to overcome was the discharging of manure into roadside ditches.

The programs certainly would have stimulated rural economies by increasing the value of farms, and creating employment for a number of professional individuals (engineers, agronomists, planners, etc.), contractors, tradesmen and construction labour. However, the opportunity to direct resources of this magnitude at such a few farms is limited. Indeed, there was indication that those owning property in watersheds adjacent to those eligible for these programs were quite discontent. The Ontario EFP model of using limited resources to influence a far greater number of farmers is more broadly applicable.

Nevertheless, it was encouraging to see a situation where large urban centres were willing to finance farm practice changes that were more for the benefit of urban dwellers than the farmer implementing the change. Of course there was a huge financial incentive for New York City to go this route in that the Watershed Agricultural Program costs taxpayers approximately $20,000 per day as opposed to the estimated $1 million dollars a day to operate a filtration plant, having already invested $6-$8 billion."

Appendix 5

AUSTRALIAN EMS STUDY TOUR

Australian Environmental Management Systems - A Canadian's Perspective - Study Tour through South Eastern Australia, October 24 – November 14, 2001

EXECUTIVE SUMMARY AND RECOMMENDATIONS, November 2001

I. "Australia has potential to capitalize on the 'Clean and Green' image through Environmental Management Systems (EMS). This could meet requirements of not only the national resource management policy but also lead to a potential marketing advantage.

II. The industry is to be congratulated for taking the initiative to launch projects, which have resulted in valuable progress, investment and momentum to generate interest in working together as an industry.

III. Farming landowners/decision-makers require clear, concise and consistent planning directives to meet on-farm targets plus local, state and national objectives.

IV. A 'Common Look' (standard format) could be one of the highest priorities at the national level with state support. Standard layout, format and design would be much easier to market to all stakeholders, including corporate sponsors.

V. Funding assistance from government could be conditional on rationalizing the format and design. Each state could custom design their variable content to mirror other states and the national framework.

VI. Plain Language expertise should also be considered with the 'Common Look'.

VII. A cost analysis to determine long-term financial investment required by the industry could be conducted on the following:

- The financial resources required by the industry for an efficiently operated EMS (several options for implementation should be compared).

- On-farm investment required for improvements to fully implement EMS requirements.

VIII. One EMS approach should not compete with another for financial assistance, or to garner support of the farmer for enrollment.

IX. Delivery options for environmental management on farms continues to be the greatest challenge (in any jurisdiction the world over) and may be a key determinant of farmer participation. Landcare groups have an extensive network of farmer members through over 4,500 local Landcare groups and would be suitably positioned for the voluntary first step introduction through a broad-based program.

X. Farmers voluntarily decide the level of rigour in reporting:

- Tier I – Voluntary as an introductory self-assessment as a first step for more rigorous audit if required.

- Tier II – Follow-up from Tier I with second party audit performed by a commodity or catchment representative to verify measures of performance.

- Tier III – Third party audit for verification and certification.

Note: Each of the above three Tiers should be built on the same format foundation.

XI. To prevent overlap or duplication, further analysis could be conducted to determine the most effective method of grass-roots "marketing" (delivery) of EMS to farmers, particularly for verification and certification: Landcare; Commodity Organization; Licensed (Private Sector) Third Party; Catchment Authority Staff.

XII. Targets within a region (particularly within a catchment), seem like a logical approach, since they are a reflection of ecological units where resources are interdependent. The question remains as to who should establish targets and what methodology is most suitable to ensure targets are realistic and achievable.

XIII. Standardized measures of performance with simple indicators need to be linked to the on-farm EMS format so that reporting can be fed into a catchment, state or national database. These are also important considerations to attract both government and corporate sponsorships.

XIV. To prevent overlap and duplication, quality assurance (QA) and food safety certification programs should be linked to the standardized EMS approach for that commodity or jurisdiction. The profile of EMS would be increased substantially.

XV. Where possible, on-farm (third party) auditors should be encouraged to become accredited in all certification schemes to increase efficiency and provide audit cost savings for the farmer."

Appendix 6

ONTARIO FARM GROUNDWATER QUALITY SURVEY 1991-1992

Members of Advisory and Working Committees

Working Committee

D. Rudolph, (WCGR) (Co- Chairman)

M. Goss (CLWS) (Co-Chairman)

A. Graham (OSCIA)

G. Kachanoski (CLWS, WCGR)

M. Scafe (MOEE)

D. Aspinall (OMAF)

B. Van den Broek (OMAF)

S. Clegg (OMAF)

D. Barry (CLWS)

J. Stimson (WCGR)

Advisory Committee

H. Rudy (OSCIA)

M. Hicknell (AAFC)

S. Singer (MOEE)

M. Brodsky (MOH)

B. Gillham (WCGR)

D. Green (OMAF)

B. Ripley (OMAF)

Appendix 7
PAST ONTARIO FORAGE MASTERS

Year	Name	County/District
2018	Doug Johnston	Perth
2016 - 2017	No provincial competition	
2015	Chris Brown	Lennox & Addington
2014	Simon Signer	Wellington
2013	Paul De Jong	Grey
2012	Thom Mueller	Ottawa-Carleton
2011	Anthony Sjaarda	Lambton
2010	Evert Veldhuizen, Jr.	Oxford (2nd place AFGC 2011)
2009	James Parson	Nipissing West
2008	Justin Dorland	Northumberland
2007	Doug Johnston	Perth
2001-2006	no provincial competition	
2000	Dave Larmer	Durham
1999	Ken Parnell	Simcoe North
1998	John Beer	Wellington (1st place, 1999 AFGC Forage Spokesperson)
1997	Barton MacLean	Lennox & Addington
1996	Brian Wiley	Grey
1995	Tom Core	Lambton
1994	Dave Woods	Oxford
1993	James McKinlay	Grey (1st place, 1994 AFGC Forage Spokesperson)
1992	John DeVries	Dundas
1991	Dan Cornwell	Oxford
1990	Brian DeJong	Durham
1989	John Ysselstein	Oxford
1988	John Markus	Oxford

Appendix 8
OSCIA PROJECTS FUNDED BY ONTARIO AGRICULTURAL ADAPTATION COUNCIL

(Source: Ontario Agricultural Adaptation Council)

Ontario Soil and Crop Improvement Association

Case No	Title	Status	Date	Dollars	Payments
AESI					
AESI-143	AESI/NSWCP Projects reports through Farm Media	Completed	01/07/2004	$17,600.00	$13,200.00
AESI-138	Pathogen Wise	Declined	/ /	$0.00	$0.00
AESI-110	Wildlife Action Project	Completed	11/30/2001	$84,100.00	$77,630.56
	3 AESI Projects				
CORDP					
8050-1	Computer Compilation of On-Farm Demonstration Trials	Completed	01/01/2000	$9,300.00	$9,300.00
	1 CORDP Projects				
CanAdapt					
CAN-405	Agricultural Plastics Recycling	Completed	06/06/2001	$29,000.00	$29,000.00
CAN-415	Campaign to Promote EFP and Stewardship Practices	Completed	11/08/2001	$115,700.00	$103,947.44
CAN-173	Compilation of Demonstration Trials	Completed	05/25/2000	$55,000.00	$55,000.00
CAN-851	Engaging the Leaders of Tomorrow - A Strategic Partnership	Declined	/ /	$0.00	$0.00
CAN-186	Feasibility Study for OSCIA Strategic	Completed	07/30/1999	$27,500.00	$26,085.00
CAN-586	Integrated Management of Emerging Field Crop Pests	Completed	08/31/2005	$221,900.00	$221,900.00
CAN-208	No-Till Corn S.T.A.R.	Withdrew	/ /	$0.00	$0.00
SPI-316	Ontario Forage Council Feasibility Study	Completed	12/21/2000	$30,000.00	$28,067.62
CAN-781	Preserving Ontario's Farmland: The Ontario Farmland Trust	Declined	/ /	$0.00	$0.00
CAN-312	Training & Outreach on Greenhouse Gas Issues in Agriculture	Completed	06/01/2001	$42,180.00	$39,765.33
CAN-4	Weather Centre	Declined	/ /	$0.00	$0.00
	11 CanAdapt Projects				
NSWCP					
NSWCP-64	American Chestnut - Species at Risk	Completed	04/01/1998	$51,800.00	$51,800.00
NSWCP-105	Enforcement of Nutrient Management	Declined	/ /	$0.00	$0.00
NSWCP-95	Feasibility Study to Provide Societal Assurance	Completed	11/01/1998	$28,000.00	$27,385.22
NSWCP-6	Ont. Nutrient Management	Completed	12/15/1997	$83,300.00	$75,544.87
NSWCP-63	Septic System Management	Completed	06/01/1998	$47,000.00	$47,000.00
	5 NSWCP Projects				
ORDP					
1093-1	Compilation of Demonstration Trials	Completed	07/01/1998	$25,100.00	$25,100.00
1237-1	Computer Compilation of On-Farm Demonstration Trials	Completed	01/01/2000	$1,618.00	$1,618.00
1128-1	Wildlife Impacts Assessment & Strategy for Agri.	Completed	11/30/1998	$200,000.00	$200,000.00
	3 ORDP Projects				

R & D III						
	8373-1	Independent Corn Adaptability Trials	Completed	05/15/2002	$26,385.00	$26,385.00
	8399-1	Wildlife Action Project	Completed	05/14/2002	$10,000.00	$10,000.00
		2 R & D III Projects				

			Total	$1,105,483.00	$1,068,729.04

25 Total Projects for Applicant*

*These projects were funded in part through federal and provincial funding initiatives delivered by the Agricultural Adaptation Council.

Appendix 9

(Source: OSCIA archives)

Beneficial Management Practices
Environmental Cost-Share Opportunities for Ontario Farmers
Available through the Canada-Ontario Environmental Farm Plan

Eligible producers may choose to participate in one or all of the federally funded environmental cost-share programs*. This brochure outlines the general categories of eligible projects, the cost share available, and the funding caps in each category. Information on program eligibility and the application process is also included.

- Canada-Ontario Farm Stewardship Program (COFSP)
- Greencover Canada (GC)
- Canada-Ontario Water Supply Expansion Program (COWSEP)

*Maximum Federal contributions per legal farm entity apply.

Delivered by
Ontario Soil and Crop
Improvement Association
(OSCIA)
April, 2005 - March 31, 2008

List of Beneficial Management Practices Environmental Cost-Share Opportunities Available through the Environmental Farm Plan 2005-2008

Beneficial Management Practices Category	Cost Share (%)	Funding Caps ($)w	Program	Beneficial Management Practices Category	Cost Share (%)	Funding Caps ($)	Program
Improved Manure Storage and Handling	30	30,000	COFS	Shelterbelt Establishment	50	10,000	GC
Manure Treatment	30	30,000	COFS	Invasive Alien Plant Species Control	50	5,000	COFSP
Manure Land Application	30	10,000	COFS	Enhancing Wildlife Habitat & Biodiversity	50	10,000	COFSP
In Barn Improvements	30	20,000	COFS	Species at Risk	50	10,000	COFSP
Farmyard and Horticultural Facilities Runoff Control	50	20,000	COFS	Preventing Wildlife Damage	30	10,000	COFSP
Relocation of Livestock Confinement and Horticultural Facilities from Riparian Zones	50	30,000	COFS	Nutrient Management Planning	50	4,000	COFSP
Wintering Site Pasture Management	50	15,000	COFS	Integrated Pest Management Planning	50	2,000	COFSP
Product and Waste Management	30	15,000	COFS	Grazing Management Planning	50	2,000	GC

Water Well Management	50	6,000	COFS	Soil Erosion and Salinity Control Planning	50	2,000	COFSP
Riparian Area Management	50	20,000	GC	Biodiversity Enhancement Planning	50	2,000	COFSP
Erosion Control Structures (Riparian)	50	20,000	GC	Irrigation Management Planning	50	2,000	COFSP
Erosion Control Structures (Non-Riparian)	50	20,000	COFS	Riparian Health Assessment	50	2,000	GC
Land Management for Soils at Risk	50	5,000	COFS	New Water Wells for Agricultural Purposes	33	5,000	COWSEP
Improved Cropping Systems	30	15,000	COFS	Ponds to Store Water for Agri. Purposes	33	5,000	COWSEP
Cover Crops	30	5,000	COFS	Spring & Sand Point Development for Agri. Purposes	33	5,000	COWSEP
Improved Pest Management	30	5,000	COFS	Water Supply to Farm for Agri. Purposes	33	5,000	COWSEP
Nutrient Recovery from Waste Water	30	20,000	COFS	Farm Water Treatment Equipment for Agri. Purposes	33	5,000	COWSEP
Irrigation Management	30	10,000	COFS	Water Supply Expansion Planning	33	5,000	COWSEP

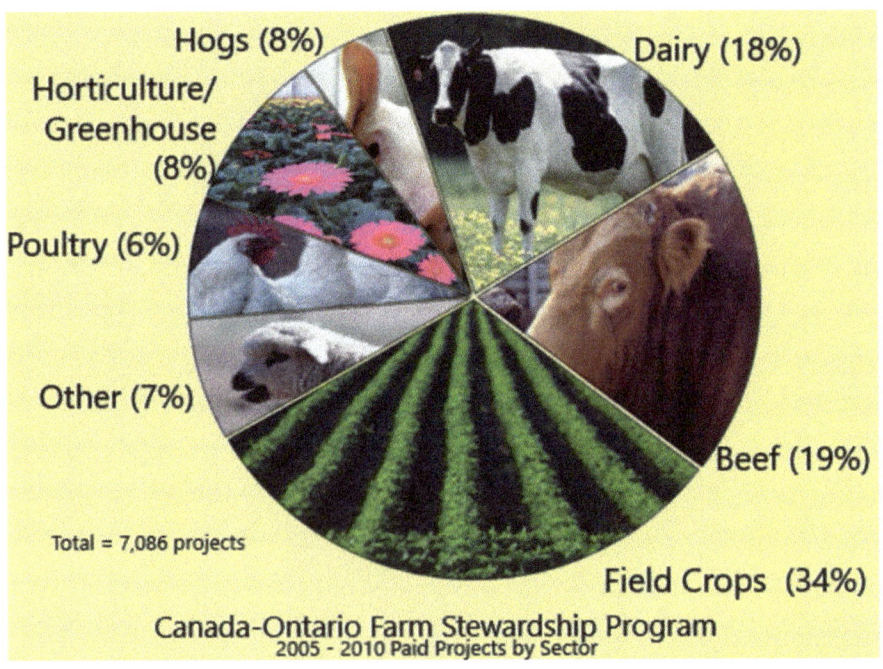

Appendix 10

SUMMARY OF OSCIA RESEARCH PROJECTS, 2006

Director's Report (Source: OSCIA archives)

Summary of the five OSCIA Special Association Projects currently under way
1. Greenhouse Gas Transition - CAPAEI
2. Optical Sensor for N application - ETAA
3. Biodiesel On-Farm
4. Cover crops' ability to Hold Nitrogen - Cord IV
5. Canada's Outdoor Farm Show

1. Canada's Agricultural Producers Addressing Environmental Issues (CAPAEI)

This is one year of bridge funding for the previous Ontario Greenhouse Gas Mitigation Program. **Projects that are taking place are:**
1. Demonstrating N Use Efficiency in Sweet Corn.
2. Horsepower, Fuel & Nitrogen Efficiencies in Manure Application Systems.
3. Assessing N Immobilization by Winter Wheat.
4. Improving N and Fertigation in Ginseng & Garlic.
5. Cover Crops for Biomass Production
6. Carbon Sequestration in Tillage and Rotation.
7. Manure injection/incorporation methods reducing contamination of water sources.
8. BMP's potential to reduce GHG Emissions.
9. Cover crops in N cycling before Vegetable crops.
10. GHG on-farm Calculator Demo.

2. Environmental Technology Assessment for Agriculture Program (ETAA)

Agriculture and Agri-Food Canada (AAFC) provided funding of $500,000 until March of 2008.

The purpose of project is :
1. Evaluate economic returns and potential to reduce N on corn.
2. Determine if "on-the-go" variable rate Nitrogen application technology is practical on-farm.

Background of technology:
Sensor emits a red or green laser light on canopy The reflectance is measured as N levels present in crop. N is applied at a variable rate according to the N Levels present in the crop on-the-go. We are trying this technology on three farm locations near Guelph, Ottawa, and Montreal.

OSCIA Fall Update for Directors
The information in this document is subject to change without notice. (OSCIA 2006)

3. Evaluating Bio-Diesel On-Farm
Natural Resources Canada provided funding of $500,000 until March of 2007.

Purpose of project is to:
- Evaluate biodiesel practicality and supply infrastructure.
- Test air quality and tailpipe emissions.
- Consumer education.

This project is being tested in: St. Clair Region (Lambton), Thames Valley Region (Middlesex), Heartland Region (Waterloo), North Eastern District (Nipissing), Quinte Region (Northumberland) and Ottawa Rideau Region (Carleton).

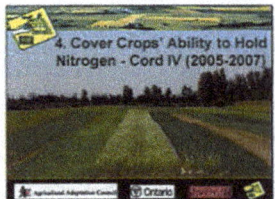
4. Cover Crops ability to hold Nitrogen over winter.
Agriculture Adaptation Council is providing $44,000 until Dec 31, 2007.
Background:
- Cover crop trials began this summer in fields that were previously growing winter wheat.
- Each site will test a number of cover crops that will be compared to a "red clover" and a "no cover" standard treatment.
- Manure and non manure treatments will be investigated to determine the ability of cover crops to trap and hold manure N for nutrient release in subsequent crops.

5. Canada's Outdoor Farm Show 2006
OSCIA and OMAFRA Crop Technology staff showcase latest in crop technology

The OSCIA/OMAFRA demonstration site is provided at no cost, and is a great chance for OSCIA to raise awareness of the organization, and talk to farmers about programs and on-farm projects.

OSCIA Fall Update for Directors
The information in this document is subject to change without notice. (OSCIA 2006)

Appendix 11
EFP First Edition Workbook – January 1994

Worksheet Contributors (Source: OSCIA archives)

1.	Soil and Site Evaluation	Bob van den Broek, OMAF (Chair)	Mike Scafe, OMEE
		Ted Taylor, OMAF	Lloyd Logan, OMEE
		Brent Kennedy, OMAF	Doug Aspinal, OMAF
2.	Water Wells	Jim Myslik, OMAF (Chair)	Doug Green, OMAF
		Mike Goss, University of Guelph	Mike Scafe, OMEE
		Tim Lotimer, Consultant, Lotowater	
3.	Pesticide Storage and Handling	Bob Stone, OMAF (Chair)	John Harvey, OMAF
		Doug Morrow, OMEE	Craig Hunter, OMAF
		Helmut Spieser, OMAF	
4.	Fertilizer Storage and Handling	Tom Sawyer, TFIO (Chair)	Henry Neutens, TFIO
		Michael Payne, OMAF	Bob Stone, OMAF
5.	Storage of Petroleum Products	Jim Myslik, OMAF (Co-Chair)	Mike Toombs, OMAF (Co-Chair)
		Sol Essop, OMEE	George Perrow, MCCR
		Bob Stone, OMAF	John Gerdels, MCCR
6.	Disposal of Farm Wastes	Hugh Fraser, OMAF (Chair)	Barb Lovell, OMAF
		Ken Hough, OCPA	Gord Coukell, Ontario Milk Marketing Board
		Dave Armitage, OFA	Garry Kay, OMEE
		Ed Barrie, OMAF	Eric Wilson, OMAF
		John Stolp, Ontario Turkey Producers' Association	
7.	Treatment of Household Wastewater	Bob Stone, OMAF (Co-Chair)	Mike Toombs, OMAF (Co-Chair)
		Dave Haymen, Upper Thames River Conservation Authority	Jim Myslik, OMAF
		Mike Bragg, MOH	Brian Cooper, OMEE

8.	Storage of Agricultural Wastes	Don Hilborn, OMAF (Chair) Peter Doris, OFAC Murray Blackie, OMEE	Norm Bird, OMAF Jack Rodenburg, OMAF Tracy Ryan, Grand River Conservation Authority
9.	Livestock Yards	Norm Bird, OMAF (Chair) Don Hilborn, OMAF John Forsyth, OMAF Murray Blackie, OMEE Peter Doris, Ontario Cattlemen's Association	Tracy Ryan, Grand River Conservation Authority Jack Rodenburg, OMAF
10.	Silage Storage	Doug Dickie, OMAF (Chair) Murray Blackie, OMEE Bob Berry, OMAF	Steve Clarke, OMAF Hank Bellman, OSA Bob Kerr, Ontario Cattlemen's Association
11.	Milking Centre Washwater	Harold House, OMAF (Chair) Harold Cuthbertson, OMAF Claude Weil, Alfred College Jack Rodenburg, OMAF	Dave Hayman, Upper Thames River Conservation Authority George MacNaughton, Ontario Milk Marketing Board
12.	Noise and Odour	Jim Myslik, OMAF (Co-Chair) Murray Blackie, OMEE	Mike Toombs, OMAF (Co-Chair) Mac Traas, Ontario Chicken Producers' Marketing Board
13.	Water Efficiency	Doug Green, OMAF (Chair) Peter Doris, Ontario Cattlemen's Association	Jim Myslik, OMAF Rick Goldt, Upper Thames Conservation Authority
14.	Energy Efficiency	Helmut Spieser. OMAF (Chair) Doug Trivers, OMAF	Ron MacDonald, Ontario Hydro John Stolp, Ontario Turkey Producers' Marketing Board
15.	Soil Management	Adam Hayes, OMAF (Chair) Peter Johnson, OMAF Ross Irwin, OFDA Don Lobb, OSCIA	Bob Stone, OMAF Doug Aspinall, OMAF Jim Eddie, OMEE

16.	Nutrient Management in Growing Crops	Keith Reid, OMAF (Chair)	Elwin Vince, OSCIA
		Mark Stauffer, Potash & Phosphate Inst.	Jim Eddie, OMEE
		John Schleihauf, OMAF	
17.	Manure Use and Management	Hugh Martin, OMAF (Chair)	Don Hilborn, OMAF
		Elbert van Donkersgood, CFFO	Ron Fleming, CCAT
		Larry Lenhardt, OCPP	Murray Blackie, OMEE
		Chris Brown, OMAF	Peter Doris, Ontario Cattlemen's Association
18.	Horticultural Production	Bill Ingratta, OMAF (Chair)	Shalin Khosla, OMAF
		Maribeth Fitts, OMAF	Jody Bodnar, OMAF
		Helen Fisher, OMAF	
19.	Field Crops Management	Neil Moore, OMAF (Chair)	Bill Curnoe, Kemptville College
		Rob Templeman, OMAF	Lisa Cruickshank, OMAF
		Tony Vyn, University of Guelph	Gilles Quesnel, OMAF
20.	Pest Control	Gord Surgeoner, University of Guelph (Chair)	Tom Hartman, OMAF
		Clarence Swanton, University of Guelph	Doug Morrow, OMEE
		Janice Schooley, OMAF	Brent Kennedy, OMAF
21.	Stream, Ditch and Floodplain Management	Bob Stone, OMAF (Chair)	John Westwood, OMEE
		Sid Vander Veen, OMAF	Jim Magee, Ontario Cattlemen's Association
		Joan McKinlay, OMAF	Donna Wales, MNR
		John Field, OMAF	
22.	Wetlands and Wildlife Ponds	Andy Graham, OSCIA (Chair)	Elizabeth Snell, Env. Consultant, Guelph
		Sid Vander Veen, OMAF	Elbert van Donkersgoed, CFFO
		Peter Roberts, OMAF	Laurie Maynard, Canadian Wildlife Service
		Ted Godowski, Ducks Unlimited	Art Timmerman, MNR
		John Westwood, OMEE	Jim Magee, Ontario Cattlemen's Association
23.	Woodlands and Wildlife	Ted Taylor, OMAF (Chair)	Ed Reid, OFAH
		John Irwin, MNR	Jim Coates, Ontario Foresters' Association
		Martin Neumann, Grand River Conservation Authority	Jim Magee, Ontario Cattlemen's Association

Abbreviations:

CCAT	Centralia College of Agricultural Technology
CFFO	Christian Farmers Federation of Ontario
MCCR	Ministry of Consumer and Commercial Relations
MNR	Ministry of Natural Resources
MOH	Ministry of Housing
OOCPA	Ontario Corn Producers' Association
OCPP	Organic Crop Producers and Processors
OFAC	Ontario Farm Animal Council
OFAH	Ontario Federation of Anglers and Hunters
OFDA	Ontario Farm Drainage Association
OMAF	Ontario Ministry of Agriculture and Food
OMEE	Ontario Ministry of Environment and Energy
OSA	Ontario Silo Association
TFIO	The Fertilizer Institute of Ontario

Grassroots Innovation
Since 1939

CPSIA information can be obtained
at www.ICGtesting.com
Printed in the USA
LVHW01s0800270918
591483LV00006B/6/P